AF378929

Coding
in
Python

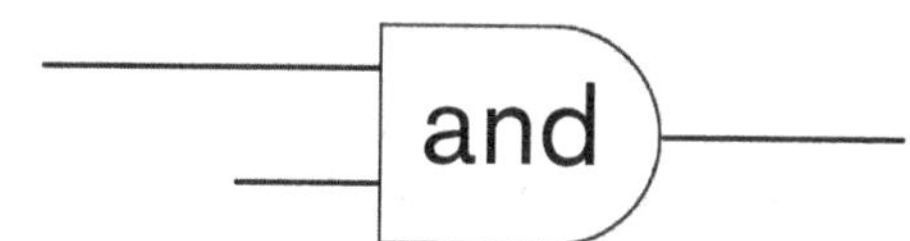

Elements of
Discrete Mathematics

Maria Litvin
Phillips Academy, Andover, Massachusetts

Gary Litvin
Skylight Software, Inc.

Skylight Publishing
Andover, Massachusetts

Skylight Publishing
9 Bartlet Street, Suite 70
Andover, MA 01810

web: http://www.skylit.com
e-mail: sales@skylit.com
          support@skylit.com

Library of Congress Control Number: 2019905086

ISBN 978-0-9972528-4-2

1 2 3 4 5 6 7   23 22 21 20 19

Printed in the United States of America

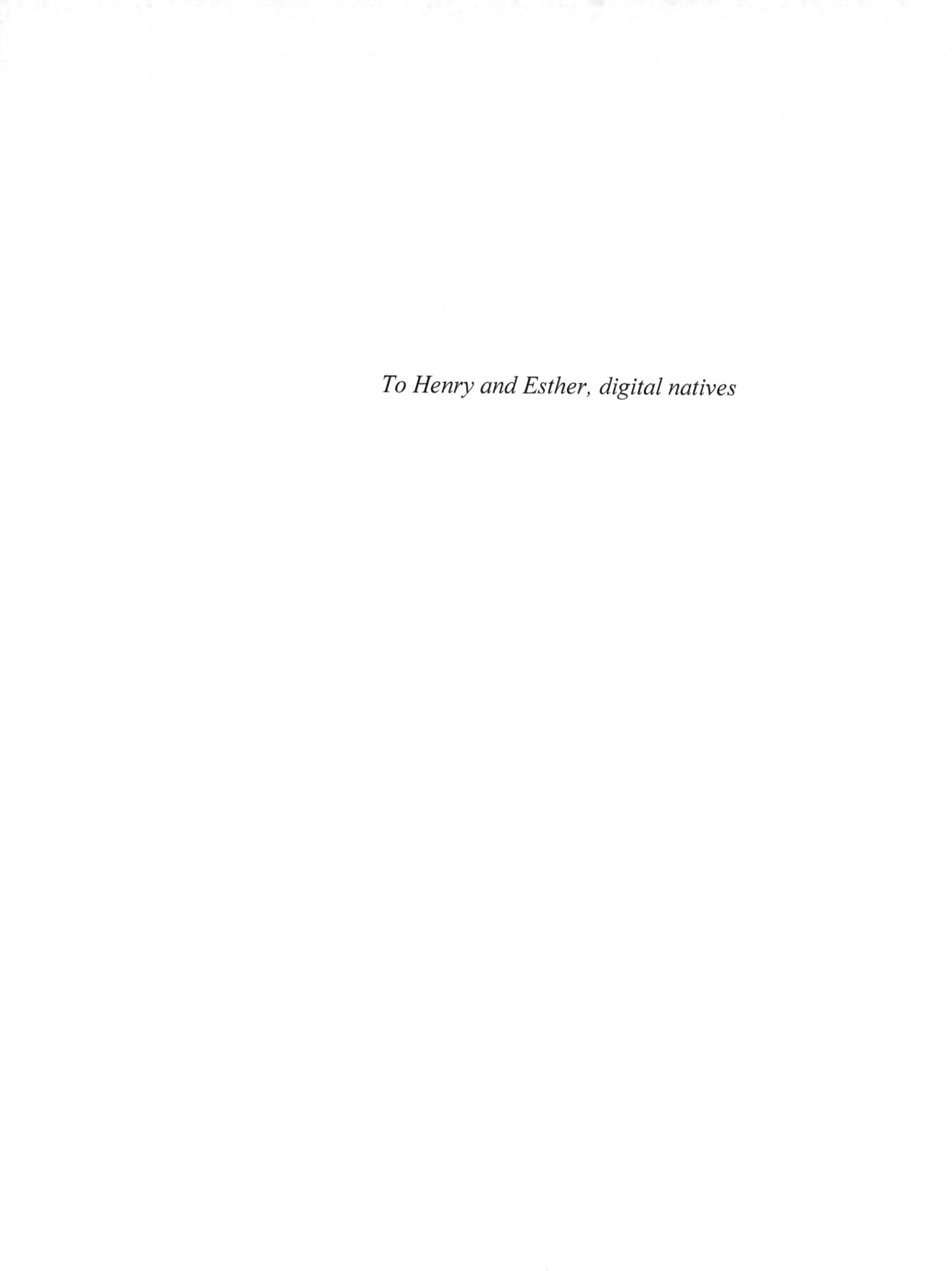

To Henry and Esther, digital natives

Brief Contents

About the Authors

Maria Litvin has taught computer science and mathematics at Phillips Academy in Andover, Massachusetts, since 1987. Prior to joining Phillips Academy, Maria taught computer science at Boston University. Maria has co-authored several popular computer science textbooks — *C++ for You++: An Introduction to Programming and Computer Science* (1998), *Java Methods: Object-Oriented Programming and Data Structures* (2001-2015), *Be Prepared for the AP Computer Science Exam in Java*, and *250 Multiple-Choice Computer Science Questions* — as well as Continental Mathematics League (CML) computer science contests for elementary and middle school students. As a consultant for the College Board, Maria provides training for high school AP Computer Science teachers, and, since 2014, as a Code.org facilitator, Maria has trained hundreds of New England elementary school teachers in teaching computer science to children in grades K-5. Maria is the recipient of the 1999 Siemens Award for Advanced Placement for Mathematics, Science, and Technology for New England and the 2003 RadioShack National Teacher Award.

Gary Litvin is the co-author of *C++ for You++, Java Methods, Be Prepared for the AP Computer Science Exam in Java, 250 MC Questions,* and CML computer science contests. Gary has worked in many areas of software development, including artificial intelligence, pattern recognition, computer graphics, and neural networks. As the founder of Skylight Software, Inc., he developed SKYLIGHTS/GX, one of the first visual programming tools for C and C++ programmers. Gary led the development of several state-of-the-art software products, including interactive touch screen development tools, OCR and handwritten character recognition systems, and credit card fraud detection software.

Contents

Chapter 5. Strings, Lists, Dictionaries, and Files

Chapter 6. Number Systems

Chapter 7. Boolean Algebra and `if-else` Statements

Chapter 8. Digital Circuits and Bitwise Operators

Chapter 9. Turtle Graphics

Preface

This book is a "Python early" remake of our earlier book *Mathematics for the Digital Age and Programming in Python*. We introduce more Python features earlier, giving the reader the necessary tools to start writing Python code sooner and in a more "pythonic" (idiomatic) manner. We have added two chapters — "Turtle Graphics" and "Vectors and Matrices" — and a separate section on Fibonacci numbers (in the "Sums and Sequences" chapter). We have updated many examples, exercises, and solutions. And we have changed the title to better match the sequence of topics and the fast-changing technological landscape and vocabulary.

But the main idea of the book remains the same: to introduce the discrete mathematics concepts that we consider essential knowledge for any literate coder. This kind of math is very accessible, yet most K-12 math curricula in the United States do not yet include it. The math segments of this book include many hands-on coding exercises, which reinforce students' learning of both coding and math.

"So, is this a math book or a computer programming book?" This is probably the first question on the impatient reader's mind. But why should it be? It is a librarian's dilemma: "Does it go on the math shelf or on the computer shelf?" There is a simple solution: put a copy on each.

The purpose of this book is to teach a particular way of thinking — precision thinking — and how to solve problems that require this way of thinking. Both mathematics and computer programming nourish the ability to think with precision and to solve problems that call for exact solutions.

Mathematics teaches us to appreciate the beauty of a rigorous argument. In the long run, this is more valuable than a lesson on solving today's practical problems. Still, mathematics does not exist in a vacuum — its abstractions are rooted in practical knowledge accumulated over centuries. The teaching of mathematics draws on examples and analogies from the world around us. At least, it should. However, the world around us is changing more and more rapidly. In the past 50 or 60 years, our world has gone digital. This change is so profound that it is sometimes hard to fully comprehend. Is that why the change remains largely ignored in our K-12 math curricula? We need to start filling the gap.

If we could build a time machine and bring Euclid over for a visit, he would find it comforting, amid the chaos of modern technologies, that geometry familiar to him is still taught in schools. Old rivals Newton and Leibniz would both find great satisfaction in the fact that tens of thousands of 11th and 12th graders are learning how to take derivatives and use integrals. But George Boole, a visitor from the more recent past, would have to search through dozens of school textbooks before he could find his algebra of propositions mentioned even in passing, even though his name is immortalized in every modern computer programming language. As for John von Neumann, a brilliant mathematician and one of the fathers of computer technology... well, with his usual optimism he would predict that within 20 years or so, every elementary school student will be learning about the AND, OR, and NOT gates. And why not?

In this book we have collected some of the easier mathematical topics that are relevant to the digital world. Many of these topics are often bundled together in freshman college courses under the name *discrete mathematics*. Discrete mathematics has become a euphemism for all elementary mathematics that is relevant today but neglected in standard middle and high school algebra, precalculus, and calculus courses. In the 1970s, Donald Knuth and his colleagues at Stanford coined the phrase "concrete mathematics" — a blend of CONtinuous and disCRETE mathematics (and also solid and not too abstract) — to describe the course Knuth taught at Stanford. Later, *Concrete Mathematics* became the title of their delightful book.[1] As they explain in their preface, Knuth "had found that there were mathematical tools missing from his repertoire; the mathematics he needed for a thorough, well-grounded understanding of computer programs was quite different from what he'd learned as a mathematics major in college."

We believe that college is too late to start. Many concepts are completely accessible to middle and high school students. And there is also another side to the relationship: just as mathematics helps achieve a deeper understanding of computer programs, some hands-on experience with computer programming helps make mathematics more tangible, more familiar, and easier to grasp.

So, if you are interested mainly in computers, we hope this book will make you a better computer programmer. If you are more interested in math, you will have ample opportunities to solve interesting problems and model some of them in computer programs. You will become familiar with fun areas of mathematics that are usually kept from middle and high school students; you will learn to solve real problems (that is, problems that you don't already know how to "solve" ahead of

---

[1] Ronald L. Graham, Donald E. Knuth, Oren Patashnik, *Concrete Mathematics: A Foundation for Computer Science*, Second Edition, Addison-Wesley, 1998.

time); you will learn the power of mathematical reasoning and proof. As a bonus, you will acquire the practical skill of programming in Python, a popular commercial programming language.

We chose Python for several reasons. First, Python gives you a chance to experiment with the language in an interactive setting with immediate feedback. Second, Python's syntax is not too complicated. Third, Python has simple yet powerful features for working with lists and "dictionaries" (maps). Finally, Python is easy to install and get started with, and it's free. Of course, there are other programming languages that have similar properties and would meet our needs. In the end, it is not any particular programming language that matters, but rather the ability to think with precision about both mathematical facts and computer programs.

This book has benefitted from the energy and precision of many friends and allies in the K-12 computer science and mathematics community.

We particularly want to thank Abby Ross of Northfield Mount Hermon high school, who read the entire book carefully and made many valuable corrections and suggestions. Dr. Patricia M. Davies of Prince Mohammad Bin Fahd University, a supporter since this book's earlier incarnation, read this version carefully and offered many useful revisions. Hans Batra of Needham High School provided helpful comments.

The previous incarnation of this book, *Mathematics for the Digital Age and Programming in Python*, benefitted from the ideas and suggestions of friends including Dr. J. Adrian Zimmer (Oklahoma School of Science and Mathematics) and Kenneth S. Oliver (formerly of Amity Regional High School in Woodbridge, Connecticut). Prof. Duncan A. Buell, then Chair of the Department of Computer Science and Engineering at University of South Carolina in Columbia, read a draft and suggested many improvements, especially for the Number Theory and Cryptology chapter.

We are grateful to Henry Garden for advice on texting habits and to Margaret Litvin for proofreading help. As ever, our deepest gratitude goes to Maria's math and computer science students: they have cheerfully test-driven the exercises presented here, provoked many good clarifications, and generally buoyed us with their enthusiasm for this material.

How to Use This Book

The *Coding in Python and Elements of Discrete Mathematics* companion web site —

> `http://www.skylit.com/python`

— is an integral part of this book. It contains downloadable student files for exercises, *Getting Started with Python* (Appendix A), links, errata, supplemental papers and syllabi, and technical support information for teachers.

Py   refers to the *Coding in Python* student (or teacher) files. For example, "See `Py\PythonCode\Fibonacci.py`" means the `Fibonacci.py` file is located in the `PythonCode` folder in `StudentFiles` (and `TeacherFiles`).

Arrow brackets like these, in the margin, mark supplementary material intended for a more inquisitive reader. This material either gives a glimpse of things to come in subsequent chapters or adds technical details.

1.■, 2.♦   In exercises, a "blue" square indicates an "intermediate" question that may require more thought or work than an "easy" question or exercise. A black diamond indicates an "advanced" question that could be treacherous, take a lot of work, or lead to unexplored territory.

✓   A checkmark at the end of a question in the exercises means that the answer or a solution is included in the student files. We have included answers and solutions to about half of the exercises. They can be found in `www.skylit.com/python/studentfiles.zip/StudentFiles/AnswersAndSolutions.pdf`.

Digital teacher files, which contain complete solutions to all the exercises and labs, are available free of charge to teachers who use this book as a textbook in their school. Go to `skylit.com/python` and click on the "Teachers' Room" link for details. A printed version is available, too.

>>> **chapter**

1

>>>

An Introduction to Computers and Coding in Python

1.1 Prologue

For a casual computer user, a computer program or an app is something that comes as an Internet download or on a CD and then runs on the computer or a smartphone. For a computer programmer, a program is a set of instructions that the computer executes to perform precisely defined tasks. Actually, computer programming is much more than just "coding" these instructions in a particular programming language, such as Python. It involves many skills, including software design, devising *algorithms*, designing *user interfaces* (screens, commands, menus, toolbars, etc.), writing and testing code, and interacting with the users of software.

In this chapter we give a quick tour of the basic computer hardware features, explain the difference between a compiler and an interpreter, and show you how to start working with Python's IDLE development environment.

1.2 CPU and Memory

At the heart of a computer is the *Central Processing Unit* (*CPU*). In a personal computer, the CPU is a *microprocessor* made from a tiny chip of silicon. The chip has millions of *transistors* etched on it. A transistor is a microscopic digital switch: it controls two states of a signal, "on" or "off," "1" or "0." The microprocessor is protected by a ceramic case less than one square inch in size, mounted on a *printed circuit board* called the *motherboard*. Also on the motherboard are memory chips and *ports* for connecting other devices (Figure 1-1).

The computer memory is a uniform series of storage units called *bytes*.

One byte holds eight bits.

One bit stores the smallest possible unit of information: "1" or "0", "true" or "false", "on" or "off".

**Figure 1-1.   Raspberry Pi 3 Model B single-board computer
fits on a board 3.5 by 2.25 inches**

The CPU can access bytes in memory in any order. This is why computer memory is called *random-access memory* (*RAM*). The same memory is used to store different types of information: numbers, letters, sounds, images, programs, and so on. All these things must be encoded, one way or another, as sequences of 0s and 1s.

A typical personal computer made in the year 2019 had 8 "gigs" (gigabytes) of RAM.

> **1** *kilobyte* (*KB*) = 1024 bytes = 2^{10} **bytes, approximately
> one thousand bytes.**
>
> **1** *megabyte* (*MB*, **"meg"**) = 1024 **kilobytes** = 2^{20} **bytes** = 1,048,576 **bytes,
> approximately one million bytes.**
>
> **1** *gigabyte* (*GB*, **"gig"**) = 1024 **megabytes** = 2^{30} **bytes** = 1,073,741,824
> **bytes, approximately one billion bytes.**
>
> **1** *terabyte* (*TB*) = 1024 **gigabytes** = 2^{40} **bytes.**
>
> **1** *petabyte* (*PB*) = 1024 **terabytes** = 2^{50} **bytes.**

A page of text with 500-600 words, without pictures or formatting, takes about three kilobytes; a high-resolution photo may take two to three megabytes, while one gigabyte can hold several hours of video in compressed MP4 format.

The CPU interprets and executes instructions stored in RAM. The CPU fetches the next instruction, interprets its operation code, and performs the appropriate operation. There are instructions for arithmetic and logical operations, for copying bytes from one location to another, and for changing the order of execution of instructions. The instructions are executed sequentially, unless a particular instruction tells the CPU to "jump" to another place in the program. *Conditional branching* instructions tell the CPU to continue with the next instruction or jump to another place depending on the result of the previous operation.

All this happens at amazing speeds. A modern CPU runs at the speed of several GHz (*gigahertz*, that is, billion *clock cycles* per second); each instruction takes one or several clock cycles.

To get a better feel for what CPU instructions are and how they are executed, let's take a look at *assembly language*, the low-level computer language that underlies the modern languages you have heard of, such as C++, Java, JavaScript, and Python.

Figure 1-2 shows a very short assembly language program for the 8088 microprocessor (which was used in the original IBM PC in the early 1980s) with the corresponding hexadecimal codes for the instructions (the *hexadecimal number system* is explained in Chapter 6) and our comments. Assembly language code is very close to the actual machine code, but it allows you to use names, rather than digits, for instructions and memory locations.

```
   Hex         Hex              Assembly              Our
 address    instruction         language            comment
               code            instruction

0AF9:0100 BB0000        MOV     BX,0000 ; move 0 into the BX register
0AF9:0103 B80100        MOV     AX,0001 ; move 1 into the AX register
0AF9:0106 3D0600    ┌─► CMP     AX,0006 ; compare AX to 6
0AF9:0109 7F05     ┌─┤─  JG     0110    ; if greater, jump to 0110
0AF9:010B 01C3     │ │   ADD     BX,AX   ; add AX to BX
0AF9:010D 40       │ │   INC     AX      ; increment AX by 1
0AF9:010E EBF6     │ └─  JMP     0106    ; jump back to 0106
0AF9:0110 90       └─► NOP             ; no operation -- skip

-g =100 0110                            ; run starting at 100

AX=0007  BX=0015 . . .
```

Figure 1-2. A segment of 8088 assembly language code

The CPU has several built-in memory units, called *registers*. The code in Figure 1-2 works with two registers, called `AX` and `BX`. For example, the first instruction — `mov bx, 0` — moves zero into the `BX` register.

We leave it to you as an exercise (Question 6) to figure out what this code computes. The result is stored in the `BX` register.

> **Mistakes in computer code are called *bugs*, and the process of eliminating mistakes in a program is called *debugging*.**

The code in Figure 1-2 was generated with the help of an ancient program called *debug*, which was supplied with the MS-DOS operating system and early versions of Windows. Debugging programs allow a programmer to execute a program step by step, in a controlled manner, and examine the contents of memory at each step. This can help find bugs when the program is not working as expected.

Section 1.2 ~ Exercises

1. Find a discarded desktop computer, <u>make sure the power cord is unplugged</u>, and remove the cover (or find online a high-resolution picture of a desktop PC with an open case). Identify the motherboard, CPU, and memory chips. Identify other components of the computer: power supply, hard disk, CD-ROM drive.

2. Computer memory is called RAM because: ✓

   (A) It provides rapid access to data.
   (B) It is mounted on the motherboard.
   (C) It is measured in megabytes.
   (D) Its bytes can be addressed in random order.
   (E) Its chips are mounted in a rectangular array.

3. My ancient PC had 512 MB of RAM and a 120 GB hard drive. How many times more storage space did the hard disk have, as compared to RAM? ✓

4. How many different values can be encoded in 2 bits? 3 bits? 1 byte?

5. ASCII (read: 'as-kee) code represents upper and lowercase letters of the English alphabet, digits, and other characters that you can find on a typical American keyboard. Each character is encoded in the same number of bits. Is one byte per character sufficient to represent all these characters? What is the smallest number of bits needed per character? ✓

6.♦ Explain the contents of the AX and BX registers after the program segment in Figure 1-2 has been executed. What does this code compute? ⸓ Hint: "hex" 15 is decimal 21. ⸓ ✓

1.3 Python Interpreter

It would be extremely tedious to write programs as sequences of digits (although in the very early days of the computer era, programmers did just that). Luckily, people quickly realized that they could define special languages for writing programs and use the computer itself to translate their programs from a high-level programming language into machine code. Some early programming languages were Fortran, COBOL, and BASIC. Some of the languages that are popular today are C++, C# ("C sharp"), Java, and JavaScript. Python is another very popular programming language; it was created in the early 1990s by Guido van Rossum at Stichting Mathematisch Centrum in the Netherlands.[*]

In a high-level programming language, each *statement* translates into several CPU instructions. Figure 1-3 shows a *function*, written in Python, that consists of a documentation string (*docstring*, a comment in triple quotes that helps to use the function) and a few statements.

A program written in a machine language or an assembly language works only on a computer with a compatible CPU. In other words, the commands/statements are specific to a particular CPU. A program written in a high-level language can be used with any CPU. For example, it can run on a PC or on a Mac.

---

[*] Contrary to popular belief, Python is not named for a snake; it is named after the British comedy group Monty Python. Their popular comedy show *Monty Python's Flying Circus* ran on the BBC from 1969 to 1974. *IDLE*, Python's *integrated development environment*, actually alludes to Eric Idle, a member of the group.

```python
def add_numbers(n):
    """Return 1 + 2 + ... + n."""
    sum1n = 0
    for k in range(1, n+1):
        sum1n += k   # add k to sum1n
    return sum1n
```

Figure 1-3. A function written in Python

There are two ways to convert a program written in a high-level programming language into machine code. The first approach is called *compiling*: a special program, called a *compiler*, examines the text of the program written in a high-level language, generates appropriate machine language instructions, and saves them in an *executable file* that is ready to run on the computer. Once a program is compiled, the compiler is not needed to run it.

The second method is called *interpreting*: a special program, called an *interpreter*, examines the text of the program, generates the appropriate instructions, and executes these instructions right away. An interpreter does not create an executable file; the interpreter is needed to run the program each time.

Compiling is like making a written translation of a text from a foreign language; interpreting is like doing a simultaneous translation while a foreign speaker is talking. An interpreter can read a program from a file, or it can allow you to enter program statements line by line, *interactively*.

Modern languages, such as Java, use a hybrid approach. First they compile a program into an intermediate low-level language, called *bytecode*, which is still independent of a particular CPU but is much more compact, closer to the machine language. Then they execute the program while interpreting its bytecode. Python, too, precompiles *modules* (libraries of functions) into bytecode.

The text of a program is governed by rather rigid *syntax rules*: you cannot just type whatever you want and expect the computer to understand it.

Every symbol in your program must be in just the right place.

In English or another natural language, you can misspell a word or omit a few punctuation marks and still produce a readable text. This is because natural languages have *redundancy*: information is transmitted with less than optimal efficiency, but this lets the reader interpret a message correctly even if it has been somewhat garbled (Figure 1-4).

Programming languages have virtually no redundancy: almost every character is essential. There are many opportunities to make a mistake, so coders have to learn patience and attention to detail, and persevere when fixing errors in the code.

Figure 1-4.   A story by Lyla Fletcher Groom, age 5
Courtesy The Writing Workshop, www.writingworkshop.com.au

❖ ❖ ❖

We are now ready to experiment with Python. Python is available for free, even for commercial applications, under the *Open Source* license. The Python license is administered by the Python Software Foundation, www.python.org/psf.

In this book, we will work with Python 3, the more recent version of Python.

The earlier version of Python is called "Python 2." There are several differences between Python 2 and Python 3 — they are not one hundred percent compatible. At the time of this writing, the latest release of Python 3 was 3.7.3.

Download the Python installer appropriate for your computer and operating system from python.org. See www.skylit.com/python, Appendix A, Getting Started with Python.

In a compiled language, you need to create the program text and save it in a file called *source code*, then run the source code file through a compiler to get an executable program. In Python, too, you can read a program from a source file. But you can also enter individual statements into the Python interpreter *shell* and see the result immediately.

It is convenient to run the Python interpreter with a GUI (<u>G</u>raphical <u>U</u>ser <u>I</u>nterface) front end. The one that comes with the standard Python installation is called *IDLE* (Figure 1-5).

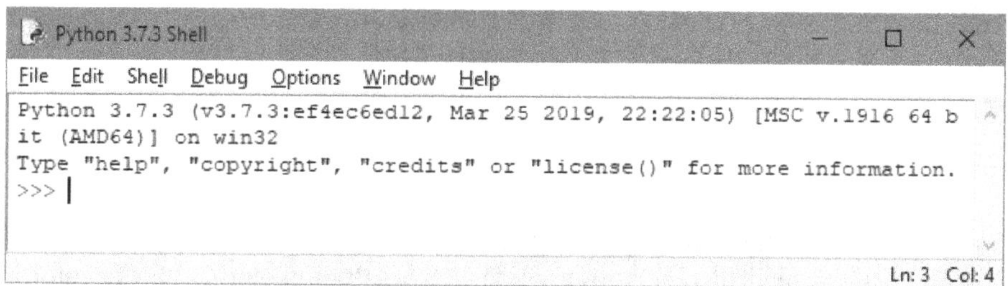

Figure 1-5.   Python's GUI "shell," called IDLE, under *Windows*

>>> is the Python interpreter's *prompt*. A prompt is a signal from a program that it is waiting for user input. The user can type in a statement and when the user presses <Enter>, the interpreter displays the result.

For example, type

```
>>> 2+3 <Enter>
```

(user input is shown in bold). Python will respond:

```
5
>>>
```

Looks reasonable! A number of things happen here. The interpreter reads the line of text (the statement) that you typed. It then analyzes the text and finds that the statement has two numbers separated by a + sign. The process of analyzing a text and extracting its components is called *parsing*.

A little experimentation will convince you that spaces do not matter (as long as the statement starts right after the prompt, with no leading spaces). For example, you can type 2 + 3 or 2 +3 or 2 +   3 — the result is the same. But if you type 2+*3, you will get

```
>>> 2+*3
        ^
SyntaxError: invalid syntax
>>>
```

Now try:

```
>>> 2(3+4)
```

You would expect 14, right? No! What you get is

```
Traceback (most recent call last):
  File "<pyshell#4>", line 1, in <module>
    2(3+4)
TypeError: 'int' object is not callable
```

Clearly the Python interpreter "thinks" there is something wrong with the statement you typed. But instead of reporting a syntax error, Python (let's call the interpreter Python for short) reports something else, which does not appear very helpful. (TypeError refers to the *type* of an object — an integer, a function, etc. — not to what you typed on the keyboard.) Apparently Python has decided that you were trying to call a function, named 2, with the input value 3+4, so Python tells you (in its own cryptic way) that 2 is not a function. You might think that Python can be really dumb sometimes. In fact, it is neither smart nor dumb — it's just a piece of code.

Meanwhile, what you really meant was

```
>>> 2*(3+4)
```

Perhaps you thought that the multiplication sign was optional, like in math. Not so.
Just as we warned you: every character matters!

Section 1.3 ~ Exercises

1.   Define *redundancy*. ✓

2.   Type `2+-3` into the Python interpreter. Is this valid syntax? Explain the
result. Now do the same for `2++3`.

3.■   Try `2+++3`. Explain the result.

4.   Try `2**3` and `2**4`. What does Python's operator `**` do?

5.   In Python, pieces of text in single or double or triple quotes represent *literal
strings*. Try `"abc" + "def"` and `'abc' + 'def'` and
`'''abc''' + '''def'''` and `"""abc""" + """def"""`. Explain what
the + operator does when it is applied to strings. ✓

6.   Can the `*` operator be applied to an integer and a string? Try `3 * 'La'` and
explain the result.

7.   Type in `9-8*2+6` and explain the result. Type in `(5-1)*(1+2)**3` and
explain the result. What is the *precedence of operators* in Python
expressions (that is, which operators are applied first)? ✓

8.■   Enter `17%3` at the Python prompt. Also try `15%4` and `15%5`. What does the
`%` operator compute? Run a few tests to determine its rank (precedence) as
compared to `+`, `-`, `*`, `/`, and `**`. ✓

1.4 Using IDLE

IDLE is a simple *IDE* (Integrated Development Environment) for editing and running Python code.[*]

A short program can be typed directly into the shell:

```
>>> s = 0
>>> for k in range(1, 7):
        s += k
        print(k, s)

1 1
2 3
3 6
4 10
5 15
6 21
>>>
```

This is not very practical, though, because you have to reenter every statement to run the program again or to make a small change in it. In IDLE, you can copy one of the previous statements: move the cursor up to the appropriate line and press `<Enter>`. You can then edit the statement.

Still, reentering every line of code would be too tedious. It is more practical to save the program statements in a file and execute the program from the file. A file that contains the text of a program is called a *source file*. Python source file names usually have the extension `.py`.

You could create a source file using any text editor, such as Notepad. You could even use a word processor — just make sure you save your file as a "text only" file, and that you replace the default `.txt` extension in the file name with `.py`. However, the easiest way to write a short Python program is by using IDLE's own built-in editor.

To open a new editor window in IDLE, choose `New File` from the `File` menu (or press `Ctrl-N`). Then type in your code. For example:

---

[*] Some people call Python a *scripting language*, because individual Python statements can be entered interactively, one by one. These people call Python programs *scripts*.

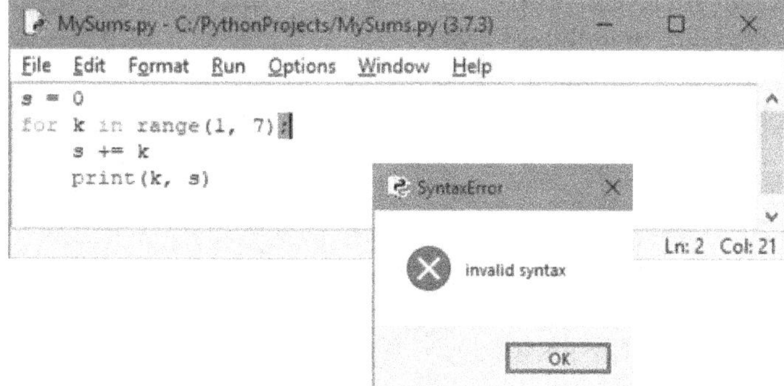

As opposed to a plain text editor, such as Notepad, the IDLE editor is "aware" of certain features of Python. For example, it highlights different elements of code in different colors. The IDLE editor automatically increases *indentation* (shift to the right) for statements that expect it: after a colon in `for`, `while`, `if`, `else`. Press `<Backspace>` to decrease the indentation level. Choose `Save As...` from the `File` menu or press `Ctrl+Shift+S` to save the program in a file. Use the `.py` extension with the file name. Save the file in a folder of your choice; for example, `C:\PythonProjects`.

While an IDLE editor window is open and active, you can test your program by choosing `Run Module` from the `Run` menu or by pressing `F5`. Python will ask you every time whether you want to save the file (click `Yes`) but you can disable this in `Options => Configure IDLE`.

If your program has syntax errors, Python will alert you to that and highlight the first error. For example:

(a semicolon was found instead of a colon).

You can have several files open at once and cut and paste text within the same window or from one window to another. Highlight the text you want to copy, press `Ctrl-C` to copy the text, position the cursor at the insertion point, and press `Ctrl-V` to paste the text.

Section 1.4 ~ Exercises

1.   The name of the programming language "Python" refers to

    (A)   the Greek letter π .
    (B)   a British comedy group.
    (C)   an elementary subatomic particle.
    (D)   a large snake, such as a boa constrictor.

2.   What is source code?

    (A)   The URL of the page that contains the program
    (B)   The text of a program in a high-level language or assembly language
    (C)   The program compiled into bytecode
    (D)   The name of the file that contains the text of the program

3.   What happens when the Python interpreter encounters a syntax error? ✓

    (A)   The interpreter continues and reports all syntax errors at the end.
    (B)   The interpreter attempts to correct the error and proceed.
    (C)   The interpreter reports the error and stops interpreting.
    (D)   The interpreter shell is closed.

4.   What happens when you enter

```
>>> import this
```

in IDLE? Try it.

5.   What happens when you enter

```
>>> import antigravity
```

in IDLE? Try it.

6.   What happens if you leave IDLE idle for a long time without typing or clicking anything? ✓

(A)   Nothing
(B)   It saves all the open files and closes all windows
(C)   It displays the message "I've been IDLE for too long!"
(D)   It displays the message "For security reasons, your session has expired. Please log in to continue."

1.5   Review

Terms introduced in this chapter:

CPU	*Assembly language*
RAM	*Compiler*
Bit	*Interpreter*
Byte	*Source code*
Kilobyte	*Syntax rules*
Megabyte	*Bug*
Gigabyte	*Debugging*
Terabyte	*Redundancy*
Petabyte	*Parsing*
Gigahertz	*Prompt*
Programming language	*Shell*

>>> chapter = 2

Variables and Arithmetic

2.1 Prologue

In this chapter we discuss some of the elements of Python code and introduce variables and arithmetic operators.

Like other programming languages, Python allows *comments* — phrases or text embedded in programs that are readable by humans but skipped by the interpreter.

Like other programming languages, Python uses a small set of *keywords* (also called *reserved words*) that have specific functions in Python programs.

The concept of a *variable* in coding is similar to a variable in algebra, but in coding it is less abstract and more practical.

Python has seven arithmetic operators: the usual +, -, *, and /, and three more that are specific to Python or coding.

2.2 Python Code Structure

In Python, the # symbol, unless it is within quotes, indicates a *comment* (Figure 2-1). The purpose of comments is to make code more readable for humans. A comment can document what a piece of code does or explain obscure code. The interpreter does not care: it simply skips all the text from # to the end of the line.

```
# This function calculates 1 + 2 + ... + n
#   using the formula sum = n(n+1)/2
def add_numbers(n):
    """Return 1 + 2 + ... + n."""
    return n*(n+1)//2 # The // operator means
                      #   integer division
```

Figure 2-1. Comments

❖   ❖   ❖

`def` and `return` in Figure 2-1 are Python's *keywords* (also known as *reserved words*) — words that have a special meaning in a programming language and should not be used for naming a programmer's own variables or functions. `def` indicates that we are defining a function; `return` specifies what value the function returns to the caller. (Once a function has been defined, you can call it from other statements. We will explain functions in more detail in Chapter 3.)

> **Note the colon at the end of the "def" line — it is required by the syntax rules.**

Python version 3.7 has 35 keywords; Figure 2-2 lists some of them.

```
import          True
from            False
as              if
                elif
def             else
return
yield           and
None            or
                not
for             in
while           is
break
continue        del
pass
```

Figure 2-2. Some of the Python keywords (reserved words)

When you enter Python code in *IDLE*, different syntactic elements are displayed in different colors. Comments are in red; keywords are in orange by default.

> **Python is case-sensitive. All reserved words must be written in lowercase letters, except the three that represent universal constants: `None`, `True`, and `False`.**

`add_numbers` is the name we gave to our function, and `n` is the name we gave to its *argument* (also called *parameter*, that is, input value).

> **It is important to give reasonably meaningful names to variables and functions.**

A name in Python may consist only of letters, digits, and the underscore characters. A name cannot start with a digit. Examples of valid names: `total`, `sum_3`, `_a`, `n2`. Python is case-sensitive, so `Total` is different from `total`. In our example, we could call our function `sumFrom1ToN`, but it is more common Python style to name variables and functions using lowercase words, separated by underscore characters, as in `add_numbers` or `sum_digits`.

❖　❖　❖

The lines that follow the "def" line are *indented* to the right. The indented lines form a block of related statements, in this case the definition of a function. Indentation must be consistent within a block — this is one of the Python syntax rules. It is customary to indent the next level by four spaces.

Some programming languages use curly braces to delineate blocks of statements. Not Python.

> **When you press <Tab> in the IDLE editor, it inserts four spaces by default, but you can change the indentation setting in IDLE's preferences.**

❖　❖　❖

In addition to comments, it is customary to include a *docstring* (documentation string) just below the `def` line (Figure 2-3). This text, often within triple quotes (which allow the comment to span multiple lines) is optional but helpful to a human reader of the code and is displayed as a "tip" when you enter the function in IDLE's interactive shell.

```
def add_numbers(n):
    """Return 1 + 2 + ... + n for
    a positive integer n.
    """                              Docstring
    return n*(n+1)//2
```

Figure 2-3. Docstring

❖　❖　❖

In Python, each statement is usually written on a separate line.

One exception is literal strings enclosed within triple quotes: ``'''`` or ``"""``. Try:

```
>>> msg = '''And it always goes on
and on and on
and on and on'''
>>> msg
```

Python echoes the value of `msg`:

```
'And it always goes on\nand on and on\nand on and on'
```

`\n` in the message signifies the *newline* character. When a string that contains `\n` is printed, `\n` shifts the next character position to the beginning of the next line. For example:

```
>>> print('And it always goes on\nand on and on\nand on and on')
And it always goes on
and on and on
and on and on
```

Or:

```
>>> print('Line 1\nLine 2')
Line 1
Line 2
```

Docstrings must use ``"""`` ... ``"""`` or ``'''`` ... ``'''`` if they exceed one line, but Python's style manual recommends always using triple quotes, even with a single-line docstring — for consistency, and to make it easier to expand.

If a statement is too long to fit on one line, you can put a `\` (backslash) at the end of the line and continue on the next line. For example:

```
>>> 1 + 2 + 3 + 4 + 5 + 6 \
   + 7 + 8 + 9 + 10
55
```

However, it is better to put the expression in parentheses instead.

The preferred style of writing an expression on multiple lines is to put it in parentheses.

For example:

```
>>> (1 + 2 + 3 + 4 + 5 + 6
 + 7 + 8 + 9 + 10)
55
```

Python lets you place several statements on one line, although this is not common style. To do that, you have to separate the statements with semicolons. For example:

```
>>> x = 3; y = 2; print(x + y)
5
```

Section 2.2 ~ Exercises

1.   Type

```
>>> from __future__ import braces
```

at an IDLE prompt (two underscore characters on each side of "future"). What is displayed?

2.   The following function returns the first character of a string or the first element of a list:

```
def first(s):
    return s[0]
```

Add a docstring to this function and type all three lines —

```
def first(s):
    < docstring >
    return s[0]
```

— at Python's prompt. Now try:

```
>>> first('Hello, world')
```

The docstring is displayed for you as a "tip" as soon as you type the opening parenthesis. Now try

```
>>> first.__doc__
```

(Python uses two underscores on each side to mark special "system" names.)

3. Which of the following are Python keywords? ✓

- ☐ (a)   `while`
- ☐ (b)   `total`
- ☐ (c)   `None`
- ☐ (d)   `define`
- ☐ (e)   `continue`
- ☐ (f)   `IF`

4. Identify four keywords in the function shown in Figure 1-3 in Chapter 1. ✓

5. Identify two <u>syntax</u> errors in the following definition of a function:

```
def bad_Code(x)
    Return x**2 - 1
```

6. What is the output from the following statement? ✓

```
print('One is better than \none' +\
    '; two is better than one')
```

7. What is the output from the following statement?

```
print('Python is #1')
```

8. Try printing the following strings.

```
print('What's up')
print("What's up")
print("She said "What's up"")
print("She said \"What\'s up\"")
```

What do \" and \' signify in *literal strings* (text within quotes)?

9.■ Find a syntax error in the following code:

```
def mystery(n):
    """Return n cubed."""
     return n**3
```
✓

2.3 Variables

In the theory of programming, a variable is a "named container" — basically a memory location with a nametag attached to it. Different values can be stored in the variable at different times. You can examine the value stored in a variable and place a new value into it.[*]

Example 1

```
>>> x = 3
>>> x
3
>>> x = 5
>>> x
5
>>> x = 2*x - 1
>>> x
9
```

Notice that x = 2*x - 1 is not an equation, as in math, but rather a set of instructions:

> Take the current value of x
> Multiply it by 2
> Subtract 1
> Store the result back in x.

In Python, every value is an *object*. Python supports several built-in types of objects: integers and real numbers, strings, lists, and so on. (Programmers can also define their own types of objects and create their *instances* in programs.) Python has a *built-in function* named type that takes any object and returns its type.

---

[*] Many "pythonistas" prefer to think of a variable name as a post-it note attached to a value.

Example 2

```
>>> x = 3
>>> type(x)
<class 'int'>
>>> x = 1.5
>>> type(x)
<class 'float'>
>>> x = 'Hello'
>>> type(x)
<class 'str'>

>>> x = '*'
>>> type(x)
<class 'str'>
```

What do we learn from these examples?

1. To introduce a new variable, all we have to do is give it a name and assign a value to it, using =.

2. If we enter the name of a variable at a prompt, Python displays its current value.

3. Python has several built-in types of objects.

    - An object of the type `int` holds an integer value.

    - An object of the type `float` holds a real number (or its approximation). The type of a real number is called `float` because such numbers are stored in the computer as *floating-point* numbers. This is discussed in Chapter 6.

    - An object of the type `str` holds a string of characters (or a single character).

4. The same variable can hold different types of objects at different times.

5. In Python each object "knows" its own type. We do not have to explicitly tell the variable what type of object is stored in it.

Names of variables are chosen by the programmer.

> **A name of a variable may consist only of letters, digits, and the underscore characters, and it cannot start with a digit. Python programming *style* calls for all names of variables to start with a lowercase letter. A name of a variable is also allowed to begin with an underscore.**

Variables can be used in *expressions*.

Example 3

```
>>> tax_rate = 0.05
>>> price = 48.00
>>> total = price * (1 + tax_rate)
>>> total
50.400000000000006
```

How does the statement

```
total = price * (1 + tax_rate)
```

work? First, Python makes sure that the variables `price` and `tax_rate` have been defined.   Then it takes their values and plugs them into the expression `price * (1 + tax_rate)`.   If one of the values is not compatible with the operation in which it is used, Python "raises an exception" (reports an error).   If everything is OK, Python calculates the result and places it into the variable `total`.

Notice that the result is not exact. The error is due to the fact that real numbers are represented in the computer approximately, as binary floating-point numbers, and there may be a small discrepancy. There is a way to get a neater output:

```
>>> print('Total sale: {0:6.2f}'.format(total))
Total sale:   50.40
```

`{0:6.2f}` is a placeholder in the format string, which right-justifies a `float` value in a field of width 6 with two digits after the decimal point; `0:` tells Python to plug in the first parameter (`total`) into the placeholder and round it appropriately.

Once an expression is evaluated and assigned to a variable, its value does not change by itself, even if you change the values of variables used in it.

Example 4

```
# continued from Example 3
...
>>> price = 60.00
>>> total
50.400000000000006
```

To see the effect of the new price, you need to recalculate the expression explicitly:

```
>>> total = price * (1 + tax_rate)
>>> total
63.0
```

A "variable" can hold a *constant*, that is, a value that will not change for the duration of the program run. For example:

```
pi = 3.1415926535897931
```

(This constant is defined in the Python module `math`. You will learn in Section 3.6 how to *import* it into a Python program.)

It is customary to write names of important universal constants in all caps. For example:

```
KM_IN_MILE = 1.60934
```

To be honest, we haven't told you the whole truth about variables. In fact, Python variables do not hold objects — they hold *references* to objects. A reference is basically the object's address in memory. Some objects, such as long strings or lists, take up a lot of space; it would be too slow to copy their values from one variable to another. Instead, when we assign `b = a`, we copy only the reference (that is, the address) from `a` into `b`. Both refer to the same object (Figure 2-4).

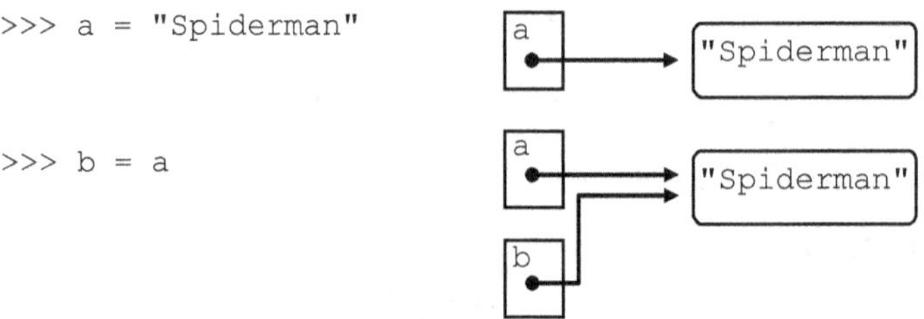

Figure 2-4. Python variables hold references to objects

This does not cause any confusion, as long as the object is *immutable*, that is, never changes. All numbers and strings are immutable objects: once created, they cannot change. But lists, for example, are not immutable. Try this:

```
>>> a = [1, 2]    # list [1, 2] is assigned to variable a
>>> b = a
>>> b
...
>>> a.append(3)
>>> b
...
```

You have to be careful!

Section 2.3 ~ Exercises

1.   Define a variable $r = 5$. Define a constant for π, accurate to five decimal places. Define a variable called `area` and set it to the area of a circle of radius r. ✓

2.   Explain the result of

```
>>> x = 5
>>> x = 2*x
>>> print(x)
```

3.■   Which of the following are syntactically valid names for variables in Python?

```
name, d7, 2x, first_name, lastName, Amt, half-price, 'Bob',
LBS_IN_KG
```

Which of them are in good style? ✓

4.   Define a variable `first_name`, equal to your first name. Define a variable `last_name` equal to your last name. Write an expression that concatenates (chains together) `first_name` and `last_name`, with a space in between, and assign that expression to a new variable.

5.   Fill in the blank:

```
>>> x = 1.0
>>> x = x + 3
>>> type(x)
```

____

6.■   What is the output from the following code?

```
a = 3
b = 2
a = a + b
b = a - b
print(a)
print(b)
```

✓

7.■   In Python,

```
name = input('What is your name? ')
```

shows the "What is your name?" prompt, then sets `name` to the string of characters typed by the user. Write a function that asks the user for his or her name and displays "Hello, *<name entered by the user>*. Welcome to Python!" Give your function a reasonable name and provide a docstring. ✓

2.4 Arithmetic Operators

As you have seen, the arithmetic operators +, -, *, and /, for addition, subtraction, multiplication, and division, work pretty much the same as in arithmetic and algebra.

Unlike in math, the multiplication sign cannot be omitted.

When parentheses are absent, multiplication and division are performed first, then addition and subtraction. Parentheses can be used the same way as in math, to modify the order of operations.

In addition to the four standard arithmetic operators +, -, *, /, Python has three more arithmetic operators:

- n%m (read "n modulo m") calculates the remainder when n is divided by m. For example, 17%3 gives 2, and 4%10 gives 4.

- The // operator is called the *integer division* operator and usually applies to integer operands. The result of n//m is the closest integer equal to or below their ratio. For example, 8//3 gives 2, and 5//9 gives 0.

- The x**n operator gives x raised to the *n*-th power. For example, 2**3 gives 8.

% and // have the same precedence (rank) as * and /; ** has higher precedence and is computed first.

Python also has a *built-in function* divmod that calculates both the quotient and the remainder when n is divided by m: q, r = divmod(n, m) sets q to n//m and r to n%m. For example:

```python
cents = 144
print(cents, 'cents =', end=' ')
quarters, cents = divmod(cents, 25)
dimes, cents = divmod(cents, 10)
nickels, cents = divmod(cents, 5)
print(quarters, 'quarters +', dimes, 'dimes +',
                nickels, 'nickels +', cents, 'cents')
```

Output:

```
144 cents = 5 quarters + 1 dimes + 1 nickels + 4 cents
```

You might have noticed in the previous examples things like `s += k`. `+=` is one of the symbols used in *augmented* (or *compound*) *assignment* statements: `+=`, `-=`, `*=`, `/=`, `%=`, `//=`, `**=`.

`x += 3` is the same as `x = x + 3`, meaning add 3 to `x`.

Other augmented assignments work in a similar way. You can use augmented assignment with any *binary operator* (that is, an operator that works with two operands). Writing things like `x = x + 1` instead of `x += 1` would be considered "uncool" in the company of seasoned coders.

`+=` also applies to two strings or lists, and `*=` applies to an integer and a string or a list.

Section 2.4 ~ Exercises

1. Remove as many parentheses as possible without changing the value of the expression and verify the result in the Python shell.

    ```
    (((3 + 7)//2)%2)**3
    ```

2. Remove all redundant parentheses from the following Python expression.
    ```
    (x - 2)**3 + (3*x)    ✓
    ```

3. What is the value of `x` after the following statements?

    ```
    x = 5
    x += 3
    x *= 5
    ```

4. Fill in the blank:

    ```
    >>> x = 5
    >>> x //= 2
    >>> x//2%2/2
    ```

5.■   Write two statements that calculate x^4 without using the `**` operator and with only two multiplications. ✓

6.■   What is the output from the following code?

```
n = 5
s = '+' * n
print(s)
```

7.   (a)   Write a function `triangle(s)` that prints an upside down triangle made of `s`, where `s` is a one-character string. For example, `triangle('*')` should print

```
***** 
 *** 
  *
```

Use three `print` statements and no other statements.

(b)■ Your function should not have `return`. Still, your function returns something. What is it and how do you know? ✓

(c)■ Write another version of `triangle(s)` that uses only one `print` statement.

(d)◆ Modify `triangle(s)` from Part (a) so that if `s` is a longer string, your function still displays a symmetrical triangle made of `s`. For example, if `s` is `'La'`, the triangle displayed should be:

```
LaLaLaLaLa 
  LaLaLa 
    La
```

≷ Hint: `len(s)` returns the number of characters in `s`; for example, `len('La')` returns 2. ≷

2.5 Review

Terms introduced in this chapter:

Comment   *Arithmetic operator*
Indentation   *Precedence of operators*
Keyword (*Reserved word*)   *Augmented arithmetic*
Variable   *statement*
Constant   *Object*

Some of the Python features introduced in this chapter:

`#` marks a comment that extends to the end of the line.
Docstring
Defining a function with `def` and returning its result with `return`
A block of related statements is indented.
Literal strings: `'abc'`, `"abc"`, `'''abc'''`, or `"""abc"""`
`\n` stands for the newline character
`\` at the end of the line means continue on the next line
`+`, `-`, `*`, `/`, `%`, `//`, `**` operators
Augmented (compound) assignments with `+=`, `-=`, `*=`, `/=`, `%=`, `//=`, `**=`
Built-in function `type`

def chapter(3):

Sets and Functions

3.1 Prologue

In mathematics, a *function* establishes a relation between a *set* of inputs (numbers, points, objects) and a set of outputs.

A function associates one output with each input.

But the same output can be associated with two or more different inputs (Figure 3-1).

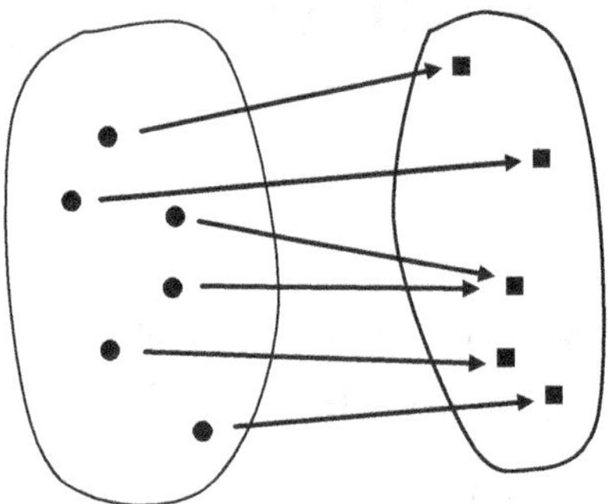

Figure 3-1. A function associates one output with each input

The concept of a function is essential for most branches of mathematics and science. A function defines a relationship between objects or quantities, and that is precisely what math and science are concerned with: how things relate to each other.

Given a function f, we often use the letter x to denote the function's input. We call the corresponding output $f(x)$. In other words, a function f maps an input x onto the output $f(x)$.

A function is also called a *mapping*: it maps the set of inputs onto the set of outputs.

An input value passed to a function is called its *argument*.

In this chapter we will discuss sets in math and in Python, different ways to define a function in math, and more advanced features of Python functions.

3.2 Sets in Math and in Python

A set of inputs... A set of outputs... You might be wondering: What exactly is a "set"? The concept of a *set* is one of those fundamental mathematical concepts that cannot be defined formally. A set is... well, any collection of unique things. For example, a set of all the students in the classroom, a set of all the letters in the alphabet, a set of all positive integers under 10.

> **The items that belong to a set are called its *elements*. A set cannot have duplicate values.**

$x \in S$ means x is an element of the set S.

Mathematicians use curly braces to list the elements of a set. For example, $A = \{1, 2, 3\}$ states that the set A contains three elements: 1, 2, and 3. The order of elements in a set does not matter.

A set can have a finite number of elements — such a set is called a *finite set*. For a finite set, we can list all its elements (although it may take some time if the set is large, such as the set of all Chinese characters). A set may consist of only one element. For convenience, mathematicians also define the *empty set* — a set that has no elements at all. The notation for the empty set is $\varnothing$.

With a little imagination, we can also define *infinite sets*. For example, a set of all positive integers, a set of all points on a line, a set of all finite sets of integers... We cannot list all the elements of an infinite set. Defining an infinite set can be tricky: infinite sets exist only in the abstract realm of mathematics. $\mathbb{N} = \{1, 2, 3, ...\}$ represents the set of all positive integers; $\mathbb{R}$ represents the set of all real numbers. We couldn't possibly list all the elements of $\mathbb{R}$ or $\mathbb{N}$.

If the set B is made up of some of the elements of the set A, then B is called a *subset* of A. The notation $B \subseteq A$ means B is a subset of A. For example, $\{2, 3\} \subseteq \{1, 2, 3\}$. For any set S, $\varnothing \subseteq S$ and $S \subseteq S$.

> **The set of all the inputs of a function *f* is called the *domain* of *f*. The set of all the outputs of *f* is called the *range* of *f*.**

If x is an element of the domain of *f*, then $f(x)$ is an element of the range of *f*.

Example 1

Let Z be the set of all integers and E the set of all even integers. The function that takes any integer n as the input and returns $2n$ as the output maps Z onto E:

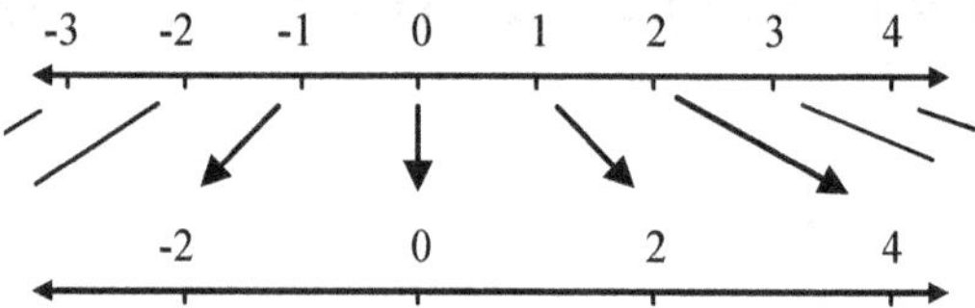

If we call this function g, we can say that $g(n) = 2n$. Z is the domain of g. E is the range of g.

Example 2

$A = \{0, 1, 2\}$ is a set of three elements. The function $0 \rightarrow 1$, $1 \rightarrow 2$, $2 \rightarrow 0$ maps A onto itself. A is both the domain and the range of this function.

❖   ❖   ❖

It is often convenient to view the range of a function as a subset of a larger set (Figure 3-2). Then we can say that a function maps its domain *into* a larger set that contains the range.

If you have two sets, A and B, you can consider all functions from A into B; for each such function its range is a subset of B (possibly equal to the whole set B).

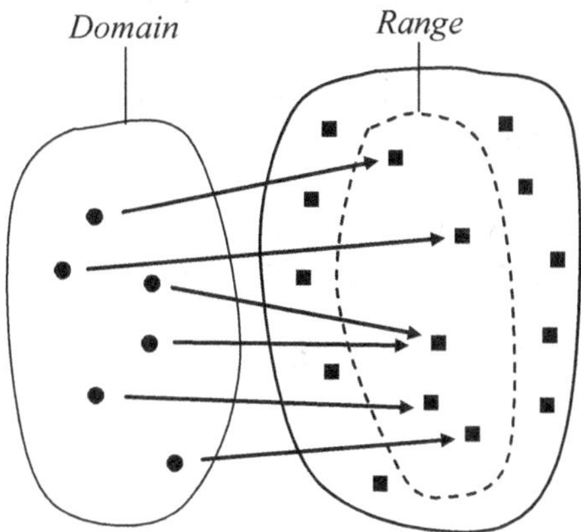

Figure 3-2. The range of a function can be a subset of a larger set

Example 3

Suppose S is the set of all the students in the classroom, and the function *birthday(p)* takes a person p as its input and returns p's birthday as the output. It is convenient to view this as a mapping from the set S (all students in the classroom) into the set D (all 366 possible birthdays, including February 29 for those special people who were born on that day in a leap year). The range of the function *birthday* — the days on which the actual birthdays of the students in the classroom happen to fall — is a subset of D.

Python represents a set in a structure that is, unsurprisingly, called `set`. The statement

```
s = set()
```

creates an empty set, named `s`. You can also create a set from a given list of values or string of characters. For example:

```
>>> countdown = set([3, 2, 1, 'Liftoff!'])
>>> countdown
{1, 2, 3, 'Liftoff!'}
>>> letters = set('abc')
>>> letters
{'a', 'c', 'b'}
```

 **The elements of a set are not arranged in any particular order.**

(When the set of positive integers is displayed in the Python shell or printed, the numbers are displayed in increasing order)

The operator `in` determines whether a given value is in the set: `x in s` gives `True` if `x` is in the set `s`; otherwise `x in s` gives `False`. For example:

```
>>> countdown = set([3, 2, 1, 'Liftoff!'])
>>> 3 in countdown
True
>>> 4 in countdown
False
>>> 'L' in countdown
False
>>> 'Liftoff!' in countdown
True
```

Python's built-in function `len(s)` returns the number of elements in the set `s`. For example:

```
>>> my_set = set([2, 3, 5, 7])
>>> len(my_set)
4
```

(The built-in function `len` also works with strings, lists, and tuples, as explained in Chapter 5.)

In Python, an object of the type `set` has a function `add` that adds an element to the set. This function is treated in the *object-oriented programming (OOP)* manner. In OOP such functions are called *methods* and are called using the object-dot syntax. For example:

```
>>> s3 = set([3, 2, 1])
>>> s3
{1, 2, 3}
>>> s3.add(0)
>>> s3
{0, 1, 2, 3}
```

If a value is already in the set, adding it again does nothing. For example:

```
>>> letters = set('Hello')
>>> letters
{'e', 'H', 'l', 'o'}
>>> letters.add('o')
>>> letters
{'e', 'H', 'l', 'o'}
```

Why do we need sets in Python? They are useful precisely because they hold values without duplicates and because they have an `in` operation that checks whether a value is in a set. `in` is implemented in a very efficient way (using a technique called *hashing*). If you misspell your username when logging into an app and it tells you "User not found," chances are the app searches a set of registered users.

> **A set in Python can hold only *immutable* objects (that is, objects that cannot change), such as numbers, strings and *tuples* (immutable lists).**

This makes sense, because in the efficient Python implementation, the object's value itself determines where the object is stored in the set. If the value were to change, the `in` operator would get confused.

Section 3.2 ~ Exercises

1. $T = \{a, b\}$ is a set of two elements. List all the different subsets of T. How many are there? (Be sure to include the empty set and the subset that is the whole set T.) ✓

2. How many different subsets does a set of three elements have? A set of four elements? A set of n elements?

3. Give an example of a function whose domain has five elements and whose range has five elements. ✓

4. Can you define a function whose domain has three elements and whose range has two elements? If yes, give an example; if not, explain why not.

5. Can you define a function whose domain has three elements and whose range has four elements? If yes, give an example; if not, explain why not.

6. Give an example of a function whose domain is the set of all integers and whose range has three elements. ✓

7. ▪   Suppose a set A has three elements and a set B has three elements. How many different functions from A into B can be defined? ✓

8.   Give an example of a function whose domain is all integers and whose range is the set of all odd integers.

9.   Devise a function that maps the set of all the points inside (and on the border) of a circle of radius 1 onto the set of real numbers $\{0 \leq y \leq 1\}$. ✓

10. ▪   Devise a function whose domain is the set of real numbers x such that $0 < x \leq 1$ and whose range is the set of real numbers y such that $y \geq 1$.

11.   What is the output from the following code segment?

```
nums = [1, 2, 3, 4, 5, 4, 3, 2, 1]
set_of_nums = set(nums)
print(len(nums), len(set_of_nums))
```

Explain. ✓

12.   Given a finite set of several non-negative integers, that is, several different integers chosen from the set $\{0, 1, 2, 3, 4, ...\}$, its *mex* is defined as the smallest non-negative number that is NOT in the set. ("mex" is short for "minimum *excludant*.") For example, *mex*($\{0, 3, 6\}$) = 1 and *mex*($\{1, 4, 5\}$) = 0. (*mex* is useful for developing optimal strategies in *combinatorial games*.)

There are 16 subsets of $\{0, 1, 2, 3\}$:

$\{\ \}$, $\{0\}$, $\{1\}$, $\{2\}$, $\{3\}$, $\{0, 1\}$, $\{0, 2\}$, $\{0, 3\}$, $\{1, 2\}$, $\{1, 3\}$, $\{2, 3\}$, $\{0, 1, 2\}$, $\{0, 1, 3\}$, $\{0, 2, 3\}$, $\{1, 2, 3\}$, $\{0, 1, 2, 3\}$

How many of these 16 sets have a *mex* of 1?

13. ▪   Fill in the blanks in this function, which calculates the *mex* of a set of non-negative numbers (as defined in the previous question).

```python
def mex(s):
    """Calculate the mex (minimum excludant) of the set s
       of non-negative integers.
    """
    n = 0
    while __________________________:


       __________________________

    return n
```

✓

3.3 Ways to Define a Function in Math

To define a function, you need to specify its domain and a way to compute the function's value (that is, the function's "output") for each element in the domain. There are several ways to define a function: in words, in a table, in a graph or a chart, or with a formula.

Example 1

A function defined in words:

"For any positive integer n, let $s(n)$ be the sum of all integers from 1 to n."

The domain of this function is all positive integers, and we can compute the value of $s(n)$ for any positive n: $s(1)=1$, $s(2)=1+2=3$, $s(3)=1+2+3=6$, and so on.

Example 2

Suppose we say something like this: "The function $f(year)$ is the number of fatal accidents in the U.S. caused by drunk drivers in a given year, for the years between 1990 and 1999." This sentence describes the function in general terms, but does not give us its values. To complete the definition, we need to look up the data and arrange it in a table and/or a graph:

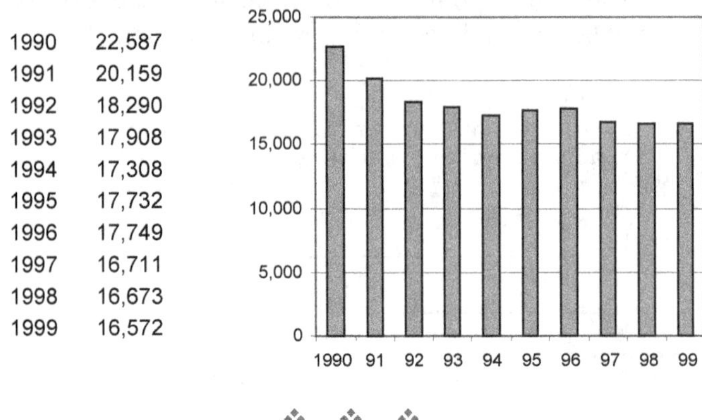

1990	22,587
1991	20,159
1992	18,290
1993	17,908
1994	17,308
1995	17,732
1996	17,749
1997	16,711
1998	16,673
1999	16,572

❖   ❖   ❖

In math and physics, by far the most common and useful way to define a function is with a formula.

Example 3

Suppose we drop a rock from a cliff. The vertical distance it travels in t seconds is given by the formula $h(t) = 16t^2$ (feet). The domain of this function is all non-negative real numbers: for any $t \geq 0$ (any time after we drop the rock), we can calculate the value of $h(t)$ from the formula. For example, $h(5) = 16 \cdot 25 = 400$ (feet). If we measure the time it takes the rock to reach the ground, we can find the height of the cliff.

❖   ❖   ❖

It is often convenient to agree ahead of time on a "universal set" on which we consider functions; for example, the set $\mathbb{R}$ of all real numbers, or the set of all points on a plane (for plane geometry), or the set of all integers (for number theory). The domains of all functions we consider are then subsets of that universal set. We also have to agree on the "universal set" to which the outputs of the functions belong.

Suppose we are working with real numbers and functions whose values are also real numbers, so both the domain and the range of a function are subsets of $\mathbb{R}$. We can write a formula, such as $f(x) = \dfrac{1}{x}$. This formula makes sense (that is, $f(x)$ can be calculated) for all $x \neq 0$. So this formula defines a function with the domain "all real numbers except 0" (or, in mathematical notation, $\mathbb{R} - \{0\}$).

If we define a function by a formula and do not specify its domain explicitly, it is assumed that the domain is the set of all numbers for which the formula makes sense.

This is called the *natural domain* of a function defined by a formula.

Example 4

For a real number x, $\sqrt{x}$ makes sense (produces a real number) if and only if $x \geq 0$. Therefore, the natural domain of the function defined by the formula $f(x) = \sqrt{x}$ is the set of all non-negative real numbers. (The range of this function is also all non-negative real numbers.)

❖   ❖   ❖

Sometimes the same function can be described in several different ways. When we prove that two definitions describe the same function, we usually obtain an interesting mathematical fact. Consider, for example, the function described in Example 1: for any positive integer n, let $s(n)$ be the sum of all integers from 1 to n, that is, $s(n) = 1 + 2 + ... + n$. Consider another function, defined by the formula $f(n) = \dfrac{n(n+1)}{2}$. In Chapter 10 we will present a proof that $s(n) = f(n)$ for any positive integer n.

Section 3.3 ~ Exercises

1.   What is the natural domain of the function defined for real numbers by the formula $f(x) = \dfrac{1}{x-2}$? What is the range of this function? ✓

2.   What is the natural domain of $f(x) = \sqrt{1+x}$? What is the range?

3.   What is the natural domain of $f(x) = \sqrt{1-x^2}$? What is the range? ✓

4.   Consider a function defined on the set of all three-digit positive integers; for each integer, the function returns the sum of its digits. For example, $f(243) = 9$. How many elements do the domain and the range of this function have? ✓

5. The table below shows a few values of the function $h(n)$ defined on the set of all positive integers:

n	1	2	3	4	5	6
$h(n)$	2	5	10	17	26	37

Come up with a simple formula that matches the function for the values shown in the table.

6. Describe a function f whose domain and range are both the set of all the points on a plane and that has exactly one *fixed point*, that is, a point P such that $f(P) = P$.

7. Given a set A of three elements, how many functions from A onto itself (that is, functions whose domain and range are both A) have no fixed points? ✓

8. Come up with a formula that defines a function on $\mathbb{R}$ with the natural domain $-1 < x < 1$ and the range $\mathbb{R}$.

9. Consider the functions f and g, defined by the following tables:

x	-1	0	1
$f(x)$	3	1	-1

x	-1	0	1
$g(x)$	-2	-1	0

What are the values of $f(g(1))$ and $g(f(1))$? ✓

3.4 Functions in Python

The concept of a function in coding is broader than in math.

A function can have more than one argument.

This is not a big deal, because several arguments can be viewed as <u>one</u>, if you group them into a single *tuple* (an immutable list of values).

Actually, mathematicians also consider functions of more than one variable. For example, $f(x,y) = \sqrt{x^2 + y^2}$.

Example 1

```python
from math import sqrt

def distance(x1, y1, x2, y2):
    """Return the distance from point (x1, y1) to point (x2, y2)."""
    return sqrt((x2-x1)*(x2-x1) + (y2-y1)*(y2-y1))

print(distance(0, 3, 4, 0))
```

The output from this program is `5.0`.

A more interesting feature is that in Python,

 **a function does not need to have a** `return` **statement.**

Even with `return` absent, a Python function returns the default value — the special constant `None`. `None` does not show in the interactive shell at all:

```python
>>> None
>>>
```

`print(None)` displays `None`:

```python
>>> print(None)
None
```

Instead of calculating a value, a function can accomplish a certain task. For example, a function can modify a list, or generate an output in a particular format, or generate a graphics display, or manipulate an image. A function that does not return a meaningful value is more like a *procedure*. Some programming languages explicitly distinguish between functions and procedures. In Python we call everything a function.

Example 2

The function below rotates the list `items` and returns `None`.

```
>>> def rotate(items):
        items.append(items.pop(0))   # Removes the first element
                             # in items and appends it at the end
>>> lst = [1, 2, 3, 4]
>>> lst
[1, 2, 3, 4]
```

When you call `rotate` in Python shell, nothing is displayed:

```
>>> rotate(lst)
>>>
```

But `lst` has changed:

```
>>> lst
[2, 3, 4, 1]
>>> rotate(lst)
>>> lst
[3, 4, 1, 2]
```

Sometimes, you might print `None` by mistake. For example:

```
>>> print(rotate(lst))
None
```

❖   ❖   ❖

A function (usually a procedure-like function) may take no arguments.

Example 3

```
>>> def print_10_dollars():
        print(10 * '$')

>>> print_10_dollars()
$$$$$$$$$$
```

Even when a function takes no arguments, you still need to put empty parentheses, both in the definition of the function and in the function call.

A Python function can return different values for the same argument, depending on user input (and in some other special situations).

Example 4

```python
def read_int():
    """Read an integer value entered by the user."""
    return int(input('Enter an integer: '))
```

The function `read_int` prompts the user to enter an integer value and returns that value.

Another example of different returns is a function that returns a random number. Python's module `random` supplies such functions, including `randint(m, n)`, which returns a random integer from `m` to `n` (inclusive):

```python
>>> from random import randint
>>> randint(1, 6)
2
>>> randint(1, 6)
1
>>> randint(1, 6)
6
>>> randint(1, 6)
4
```

A Python function can also return several values. Actually, Python can automatically pack several values into a *tuple* (immutable list) and automatically unpack the tuple and assign individual values to variables. For example:

```python
>>> t = 2, 3   # t is a tuple
>>> t
(2, 3)
>>> t2 = ('hi', 'bye', 'there')   # t2 is a tuple
>>> t2
('hi', 'bye', 'there')
>>> x, y = t
>>> x
2
>>> y
3
```

You can also swap two values in one statement:

```
>>> a, b = b, a
```

The `divmod` function that we mentioned in Chapter 2 returns a tuple consisting of two values: `a//b` and `a%b`.

To summarize, a Python function may take one or more arguments, take no arguments at all, return `None`, or return one or several values (packed in a tuple), and can even return different values given the same input value, depending on user input or other circumstances. Thus Python stretches the mathematical concept of a function beyond recognition.

Python also stretches the concept of the domain of a function. In Python, the domain of a function is basically the set of all objects for which the function works. If the function is called with an argument outside its domain, it *raises an exception*.

Example 5

```
def difference_set(s1, s2):
    """Return the set of distinct values from s1 that are not in s2."""
    diff = set()
    for e in s1:  # for each element in s1
        if e not in s2:  # if that element is not in set s2
            diff.add(e)
    return diff
```

This function works for any *iterable* sequences `s1` and `s2`: strings, lists, tuples, sets. `s1` and `s2` can even be of different types.

```
>>> difference_set('We love Python', ['e', 'o', 'y', ' '])
{'l', 'P', 'W', 't', 'v', 'h', 'n'}
```

However, `difference_set` raises an exception (reports an error) if `s1` or `s2` is not iterable:

```
>>> difference_set('We love Python', 2)
...
TypeError: argument of type 'int' is not iterable
```

So, for all practical purposes, we can view a Python function simply as a fragment of code that is callable from other places in a program.

> **Using functions helps you structure your code better and avoid duplicate code when you need to perform the same task or calculation several times.**

A function may take inputs (arguments), perform a certain task or calculation, and return a value to the caller. Or it may raise an exception if the conditions for performing the task are not right.

Section 3.4 ~ Exercises

1. Write a function that concatenates two given strings, with a space between them, and returns the new string. ✓

2. Write a function `print_house` that prints

    Can you do it using only one `print` call?

3. Pretend that string's `rjust` method does not exist and write your own:

```
def right_justify(s, w):
    """Pad the given string s with spaces on the left to
        form a new string of total length w and return
        the new string.
    """
    return ________________________________
```

    ⌇ Hint: Recall that `len(s)` returns the length of string s. ⌇

    Try

```
>>> right_justify('123', 5)
>>> right_justify(5, '123')
>>> right_justify('123')
>>> right_justify('12345', 3)
```
    ✓

4. Type in the function `print_10_dollars()` from Example 3 into the Python shell, then try

```
>>> print(print_10_dollars())
```

Explain the output. Define another function, `make_10_dollars`, that prints nothing but <u>returns</u> `'$$$$$$$$$$'`. Explain the output of

```
>>> print(make_10_dollars())
```

5. Consider the following function:

```
def print_triangle(n, ch):
    while n > 0:
        print(n * ch)
        n -= 1
    return 1
```

Try to predict the output from

```
>>> print_triangle(2, '*') + print_triangle(3, '#')
```

✓

6. Suppose we define the function `add_numbers` as follows:

```
def add_numbers(n):
    if n > 0:
        return n*(n+1)//2
```

Try

```
>>> print(add_numbers(6))
```

and

```
>>> print(add_numbers(0))
```

Explain the results.

7. Call the function `read_int` from Example 4, and when it prompts you to enter an integer, just press <Enter> (or type "asdfgh" and press <Enter>). Explain the result.

8.   Which of the following calls return a result and which raise an exception? ✓

- ❑ (a)   `len(0)`
- ❑ (b)   `len('0')`
- ❑ (c)   `len('''0''')`
- ❑ (d)   `len('''''')`
- ❑ (e)   `len([0])`
- ❑ (f)   `len([])`
- ❑ (g)   `len((0, 1))`
- ❑ (h)   `len(range(0, 1))`

3.5 Function Arguments

The variables that stand for input values in a definition of a function are called *formal parameters*, but people often just refer to them as parameters or arguments.

Example 1

```python
from math import sqrt

def distance(x1, y1, x2, y2):
    """Return the distance between the points (x1, y1) and (x2, y2)"""
    return sqrt((x2-x1)*(x2-x1) + (y2-y1)*(y2-y1))
```

`x1, x2, y1, y2` are the formal parameters of the function `distance`.

To call this function we have to pass four arguments to it. For example:

```python
>>> d = distance(0, 3, 4, 0)
```

> **When a function is called, the number and order of arguments passed to it must match the number and order of the parameters that the function expects.**

Objects passed to the function as arguments are copied into the function's formal parameters.

More precisely, the references to (addresses of) the objects passed to the function are copied into the formal parameters.

Example 2

```
>>> def distance(x1, y1, x2, y2):
        return sqrt((x2-x1)*(x2-x1) + (y2-y1)*(y2-y1))
>>> a1=0
>>> b1=3
>>> a2=4
>>> b2=0
>>> distance(a1, b1, a2, b2)
5.0
```

❖   ❖   ❖

Within the function, formal parameters act as *local variables*. Their *scope* is (that is, they exist only) within the function's definition, from the place where the variable is defined to the end of the function's code.

There is nothing wrong with reusing the names of the formal parameters as the names of the actual arguments passed to the function.

Example 3

```
>>> def distance(x1, y1, x2, y2):
        return sqrt((x2-x1)*(x2-x1) + (y2-y1)*(y2-y1))
>>>
>>> x1=0
>>> y1=3
>>> x2=4
>>> y2=0
>>> distance(x1, y1, x2, y2)
5.0
```

❖   ❖   ❖

Python lets you supply default values for some of a function's arguments. The arguments with default values must appear at the end of the list in the `def` statement (or use parameter name = value.) For example:

```
def distance(x1, y1, x2=0, y2=0):
    """Return the distance between the points (x1, y1)
        and (x2, y2).
    """
    return sqrt((x2-x1)*(x2-x1) + (y2-y1)*(y2-y1))
```

Now, instead of `distance(x, y, 0, 0)`, you can simply call `distance(x, y)`.

> **If you pass an immutable object to a function, the function cannot change it.**

Example 4

```
>>> def nice_try(msg):
        msg += '***'

>>> msg = "You can't change me!"
>>> msg
"You can't change me!"
>>> nice_try(msg)
>>> msg
"You can't change me!"
```

The `msg` you changed inside the `nice_try` function is <u>not</u> the same as the variable `msg` outside the function — `nice_try`'s `msg` is a <u>copy</u> of `msg`. So the statement `msg += '***'` creates a new string and places a reference to that string into the formal parameter `msg`, which acts like a local variable inside the `nice_try` function. That new value of `msg` is never used: when the function is exited, its local variables are destroyed. The *global variable* `msg` remains unchanged. To make the function work as intended, it should <u>return</u> the new string:

```
def append_stars(msg):
    return msg + '***'
```

```
>>> msg = 'Yes, you can change me'
>>> msg
'Yes, you can change me'
>>> msg = append_stars(msg)
>>> msg
'Yes, you can change me***'
```

Section 3.5 ~ Exercises

1.   Write and test a function that takes two values, *a* and *b*, and returns their *arithmetic mean*, $\dfrac{a+b}{2}$.

2.   What is the result of the following dialog?

```
>>> def swap(x, y):
        x, y = y, x
>>> a = 1
>>> b = 2
>>> swap(a, b)
>>> a
...
>>> b
...
```

Explain. ✓

3.   (a) Recall that the quadratic equation $ax^2 + bx + c = 0$ may have two roots:

$$x_1 = \frac{-b + \sqrt{b^2 - 4ac}}{2a} \text{ and } x_2 = \frac{-b - \sqrt{b^2 - 4ac}}{2a}.$$ Write and test a function

that takes a, b, and c as arguments and returns two values that are the values of the two roots. Calculate $\sqrt{b^2 - 4ac}$ only once. Leave it to sqrt to raise an exception when $b^2 - 4ac < 0$. ⸵ Hint: don't forget

```
from math import sqrt
```

⸵ ✓

(b) Modify your function from Part (a) so that it returns None, None if $b^2 - 4ac$ is negative.

(c) Write a program that prompts the user to enter three real numbers, a, b, and c, calls the function from Part (b), and displays the result. If the function returns None, None, display "No solutions in real numbers".

4.♦   Review the explanation of default values for function arguments. What are Python's responses in the following dialog?

```
>>> def double(x=0):
        return 2*x

>>> double
...
>>> double()
...
```

5.■   The function below prints obj *n* times.

```
def print_n_times(n, obj):
    """Print obj n times."""
    print(n*str(obj))
```

Type in this definition and try

```
>>> print_n_times(5)
```

What happens? Change the function to print *n* stars ('*' characters) when the obj argument is not supplied.

3.6 Python's Built-In Functions

Python has a number of *built-in* functions that make coding in Python easier.

`max` and `min`

`max` and `min` work with any *iterable* sequence (list, tuple, string, range, set) of numbers or strings. They also work with a few values, separated by commas.

```
>>> x = 2.5; y = 3
>>> max(x, y)
3
>>> t = (4, 1.3, 2.5)
>>> min(t)
1.3
>>> max('We love Python')
'y'
>>> max(range(100))
99
```

`len` and `sum`

`len` works with any iterable sequence (list, tuple, string, range, set) and returns the number of elements in it.

```
>>> len('ABC')
3
>>> len(range(100))
100
```

`sum` works with any iterable sequence of numbers.

```
>>> sum(range(1, 10, 2))
25
```

(`range(1, 10, 2)` generates the sequence 1, 3, 5, 7, 9.)

```
>>> sum([1, 1, 2, 3, 5, 8])
20
>>> sum(x*x for x in range(1, 11))
385
```

($1 + 4 + 9 + 25 + 36 + 49 + 64 + 81 + 100 = 385$.)

`abs, pow, round, divmod`

`abs(x)` returns the absolute value of `x`.

`pow(x, y)` returns x^y, the same as `x**y`. `pow(x, n, p)` returns `(x**n) % p`.

`round(x)` returns `x` rounded to the nearest integer. `round(x, k)` returns `x` rounded to `k` digits after the decimal point.

`divmod(a, b)` returns the pair `a//b, a%b`.

`input` and `print`

`input(msg)` displays `msg` as prompt and returns the string entered by the user.

`print(...)` prints several values, separated by spaces. An optional parameter `end=s` prints the string `s` instead of newline at the end. For example:

```
>>> print(2, end='+'); print(3, end='='); print(5)
2+3=5
```

<u>int, float, str</u>

These functions convert their argument into the corresponding type:

```
>>> int('456')
456
>>> str(456)
'456'
>>> float(1)
1.0
```

<u>set, list, tuple, sorted, reversed</u>

set, list, and tuple convert the sequence received as their parameter to a set, list, and tuple, respectively.

```
>>> list('abc')
['a', 'b', 'c']
>>> tuple('abc')
('a', 'b', 'c')
>>> tuple([1, 2, 3])
(1, 2, 3)
```

sorted converts a sequence of numbers or strings into a list, arranged in ascending order. You can also sort the sequence in descending order, if you supply the optional argument reversed=True.

```
>>> sorted([1, 5, 2, 4, 3])
[1, 2, 3, 4, 5]
>>> sorted([1, 5, 2, 4, 3], reverse=True)
[5, 4, 3, 2, 1]
>>> sorted('dcba')
['a', 'b', 'c', 'd']
```

reversed is used as an *iterator*:

```
>>> for x in reversed([1, 2, 3, 4, 5]):
        print(x)
5
4
3
2
1
```

> **To get an explanation of any built-in function, type**
> **help(*function name*) at the Python prompt.**

> **Many built-in functions — especially len, sum, max, min, list, str,**
> **pow — have short, expressive names. You will be tempted to use the**
> **same names for your own variables. Don't! If you do, you will disable**
> **the function and get errors if you try using it.**

We will discuss other built-in functions in later chapters. You can find a list of more frequently used built-in functions in Appendix B, and the complete list in https://docs.python.org/3/library/functions.html.

In addition to built-in functions, Python's math module includes essential math functions: sqrt, log, sin, cos, factorial, and so on, and the constants pi and e. To use these functions and constants you need to *import* them into your program. For example:

```
from math import sqrt, log, pi
```

To see the complete list of math functions, type

```
>>> import math
>>> help(math)
```

or see https://docs.python.org/3/library/math.html.

Section 3.6 ~ Exercises

1.   What do min(range(n)) and max(range(n)) return when n is a positive integer? ✓

2.   Using the built-in functions str and len, write a one-line function num_digits that returns the number of digits in a given positive integer.

3.◆   Using the built-in functions str, int, and sum, write a one-line function that returns the sum of the digits in a given positive integer. ✓

4. Create an empty set `s` and try to add the list `[1, 2, 3]` to it as an element:

```
>>> s = set()
>>> s.add([1, 2, 3])
```

What happens? Now try to add the tuple `(1, 2, 3)`:

```
>>> s.add((1, 2, 3))
>>> len(s)
...
```

What is the result?

5. What is the result in the following dialog?

```
>>> x = input('Enter the first number: ')
Enter the first number: 2
>>> y = input('Enter the second number: ')
Enter the second number: 3
>>> print(x + y)
...
```

Correct the above statements to print 5.

6. What does `pow(2, 8, 5)` return?

7. What is the output from the following statements?

```
from math import sqrt
print(sorted(set([round(sqrt(n)) for n in range(1, 20)])))
```

✓

3.7 Review

Terms and notation introduced in this chapter:

Set	*Function argument*	$A = \{a, b, c\}$
Element of a set	*Natural domain*	$x \in A$
Finite set	*Object-oriented*	$\varnothing$
Infinite set	*programming (OOP)*	$A \subseteq B$
Subset	*Method*	$f(x)$
Empty set	*Mutable / immutable*	
Function	*objects*	
Mapping	*Formal parameter*	
Domain	*Procedure*	
Range		

Some of the Python features mentioned in this chapter:

```
s = set()
s = set([3, 5, 8])
return
None
def somefun(..., x=<default value>):
```

Built-in functions: `len, sum, max, min, abs, round, pow, divmod, input, print, set, list, tuple, sorted, reversed`

```
from math import sqrt, pi, log

from random import randint

>>> help(<function name>)
>>> import math; help(math)
>>> import random; help(random)
```

while chapter is 4:

Algorithms and

while and for Loops

4.1 Prologue

An *algorithm* is a precise description of the steps necessary to accomplish a task or calculation.   In this chapter we will show you how to describe and compare algorithms.

The `while` and `for` loops in Python are used to implement *iterations* (that is, repeating a block of statements several times). These tools are essential for coding "interesting" algorithms.

4.2 Algorithms

> **An *algorithm* is a precise description of the steps necessary to calculate the value of a function or to accomplish a task.**

A recipe in a cookbook is a type of algorithm. But "interesting" algorithms usually involve several steps that are repeated many times. This allows us to "fold" a long sequence of steps into a relatively short algorithm — which gives algorithms their power.

A good algorithm has three properties:

- A good algorithm is <u>compact</u>: the running times may be different for different sizes or parameters of the task, but the length of the description of the algorithm and the code of a program based on it remain the same.

- A good algorithm is <u>general</u>: it works for different sizes or parameters of the task.

- A good algorithm is <u>abstract</u>: it does not depend on a particular programming language or computer system.

Example 1

The following algorithm calculates $1-\dfrac{1}{3}+\dfrac{1}{5}-...+\dfrac{1}{9997}-\dfrac{1}{9999}$. (Four times this sum approximates π with some degree of accuracy. Adding more terms to the sum produces a better approximation.)

> Set *k* to 1
> Set *sign* to 1
> Set *sum* to 0
> While *k* < 10000, repeat the following steps:
> > Add *sign* / *k* to *sum*
> > Add 2 to *k*
> > Set *sign* to –*sign*
> Return *sum*

This description is compact and abstract, but it is not general, because it works only for a particular parameter: 10000. There is no need to "hardcode" 10000 into the algorithm — it works for any positive integer *n*. We can replace 10000 with *n*. *n* can be entered by the user or come from another place in the program.

❖   ❖   ❖

Algorithms are often described not in words but in a more concise notation, called *pseudocode*. In pseudocode the above algorithm might look like this:

```
Input: n
k ← 1
sign ← 1
sum ← 0
while k < n:
    sum ← sum + sign/k
    k ← k + 2
    sign ← -sign
Output: sum
```

Pseudocode is not a programming language, so the algorithm above is not real computer "code." But you can *implement* this pseudocode algorithm in a programming language of your choice.

Another way to describe an algorithm is with a *flowchart*. A flowchart for the above algorithm might look like this:

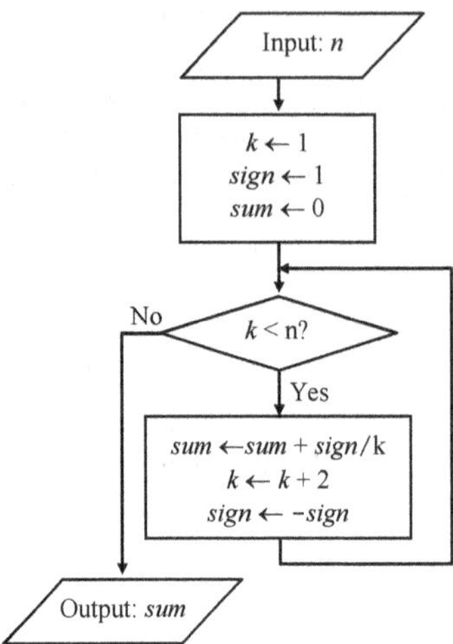

Flowcharts belong in a computer museum — they are hardly used nowadays.

Section 4.2 ~ Exercises

1.   Name a couple of algorithms that you know from arithmetic and algebra.

2.   Devise an algorithm for calculating $s(n) = 1 + 2 + ... + n$ without using the formula $s(n) = \dfrac{n(n+1)}{2}$. Present your algorithm in pseudocode. ✓

3.   Suppose you only have the operations of addition and subtraction. Devise an algorithm that takes two positive integers, m and n, and calculates the quotient and the remainder when m is divided by n. (For example, when 17 is divided by 5, the quotient is 3 and the remainder is 2.) Present your algorithm in pseudocode. ✓

4.   Devise an algorithm for calculating 3^n, where n is a non-negative integer. Write your algorithm in pseudocode.

4.3 `while` and `for` Loops

In this section we will create a Python program that prompts the user to enter a positive integer n_max and prints out the sums $1+2+...+n$ for n from 1 to n_max. But first let us briefly review what we have learned so far about Python's arithmetic operators.

Python supports `int` and `float` types of numbers. (It also supports the `complex` type, but we will leave this alone for now.) Python has seven arithmetic operators: `+`, `-`, `*`, `/`, `//`, `%`, and `**`.

When `+`, `-`, and `*` are applied to two `int`s, the result is an `int`. If at least one of the operands is a `float`, the result is a `float`. The division operator `/`, when applied to two integers, gives a `float`.

Python's integer division operator `//`, when applied to two integers, `a` and `b`, gives the largest integer that does not exceed a/b. For example, `15//2` gives `7`.

For your convenience, Python also supports *augmented* (also called *compound*) *assignment* statements, using `+=`, `-=`, `*=`, and so on. `a += b` is the same as `a = a + b`; `a -= b` is the same as `a = a - b`, and so on.

❖ ❖ ❖

Our program should print n and the sum $1+2+...+n$ for all n from 1 to a certain number n_max entered by the user. But we can't just write

```
print(1, 1)   # print 1 and 1
print(2, 3)   # print 2 and 1 + 2 = 3
print(3, 6)   # print 3 and 1 + 2 + 3 = 6
print(4, 10)  # print 4 and 1 + 2 + 3 + 4 = 10
...
```

because we don't know ahead of time what n_max will be. It would also make little sense, since we would have to compute all the sums by hand. Even if we knew n_max, say n_max = 1000, and used a formula for the sum, our program would be too long:

```
n = 1
print(n, n*(n+1)//2)
n = n + 1
print(n, n*(n+1)//2)
n = n + 1
print(n, n*(n+1)//2)
n = n + 1
... # 2000 lines of code
```

Fortunately, Python provides an *iterative statement*, called a *while loop*, which allows us to <u>repeat</u> the same block of statements multiple times while a certain condition holds true. The syntax for the `while` loop is:

```
while <condition>:
    ...
    ...
```

<*condition*> is an expression that can use the *relational operators* <, >, <=, >=, == or `is` (equals), and `!=` or `is not` (not equals) or the operator `in` (an element of). The value of <*condition*> is either `True` or `False`. Try it:

```
>>> 6 <= 6
True
>>> 5 <= 6
True
>>> 5 <= 4
False
```

<*condition*> can also include *logical operators*: `and, or, not`. For example:

```
>>> x = 4
>>> x > 0 and x <= 3
False
>>> not x < 0
True
```

As long as the condition remains true, the program repeats the statements in the `while` block (the statements that are indented under `while`). Such repetitions are called *iterations*. Usually the condition includes a test for a variable that is updated on each iteration, so eventually the condition becomes false and the iterations stop. The program then continues with the statement that follows the `while` block.

Example 1

```python
n_max = 10
n = 1
while n <= n_max:
    print(n, n*(n+1)//2)
    n += 1
```

The output is:

```
1 1
2 3
3 6
4 10
5 15
6 21
7 28
8 36
9 45
10 55
```

We can make the output prettier by using string's `format` function. For example:

```python
print('{0:4d} {1:10d}'.format(n, n*(n+1)//2))
```

Or, by using a so-called *f-string*:

```python
print(f'{n:4d} {n*(n+1)//2:10d}')
```

The output will then be

```
   1           1
   2           3
   3           6
   4          10
   5          15
   6          21
   7          28
   8          36
   9          45
  10          55
```

Thanks to the loop, the program's code has the same length regardless of whether we run it with `n_max = 10` or `n_max = 10000`. The output is different, of course.

Instead of using the `while` loop, we can use a `for...in range` loop.

Python's `range` is an object that generates a sequence of uniformly spaced numbers. (In mathematics, such a sequence is called an *arithmetic sequence*.) `range` is quite flexible:

`range` call	Sequence of numbers generated
`range(5)`	0, 1, 2, 3, 4
`range(n)`	0, 1, 2, ..., $n-1$ (n is not included)
`range(3, n)`	3, 4, ..., $n-1$
`range(0, 10, 2)`	0, 2, 4, 6, 8
`range(2, 9, 3)`	2, 5, 8
`range(9, 2, -1)`	9, 8, 7, 6, 5, 4, 3 (2 is not included)

Our program becomes shorter and more readable with `for...in range`:

```
n_max = 10
for n in range(1, n_max+1):   # n_max+1 to include n_max
    print(n, n*(n+1)//2)
```

Example 2

Suppose we don't know the formula for the sum and want to use a "brute force" approach. We can calculate the sum $1+2+...+n$ using a loop:

```
sum_n = 0
for k in range(1, n+1):
    sum_n += k
```

This loop can be *nested* within the first loop:

```
n_max = 10
for n in range(1, n_max+1):     # +1 to include n_max
    sum_n = 0
    for k in range(1, n+1):
        sum_n += k   # same as sum_n = sum_n + k
    print(n, sum_n)
```

Example 3

The above version of the program works, but it is inefficient because it keeps recalculating sums from scratch, over and over. If we have found $1 + 2 + ... + 49$, we only need to add 50 to it to get $1 + 2 + ... + 49 + 50$.

We can keep track of the sum and eliminate the nested loop, like this:

```
n_max = 10
sum_n = 0
for n in range(1, n_max+1):
    sum_n += n
    print(n, sum_n)
```

You might say: "Who cares? Computers are fast anyway." However, such inefficiencies can visibly slow down even the fastest computer if the task is large. Here the version with nested loops will perform a number of operations roughly proportional to n_max^2; in the streamlined version with one loop, the number of operations will be proportional to n_max — a big difference. If n_max is, say, 1,000,000, the efficient program will run fine (the output will be quite long, though) but the program with nested loops will work for ten hours. In this example, the more efficient code is also slightly shorter.

❖   ❖   ❖

We have almost completed what we set out to do. The only remaining question is: How do we get n_max from the user?

As we know, Python has the built-in function `input`, which lets you issue a prompt and read a string from the keyboard.

Example 4

```
>>> word = input('Enter any word: ')
Enter any word: Hello
>>> word
'Hello'
```

`input` returns a string. We can convert it into an `int` by applying the built-in function `int`:

```
>>> n_max = int(input('Enter a positive integer: '))
Enter a positive integer: 10
>>> n_max
10
```

This works, as long as the user enters a string that represents a valid positive integer. But what if the user mistypes? For example:

```
>>> n_max = int(input('Enter a positive integer: '))
Enter a positive integer: 1.0
...
ValueError: invalid literal for int() with base 10: '1.0'
```

The program is aborted with a rather cryptic error message. In the technical Python lingo, the program *raises an exception* — in this case, a ValueError exception. This is not very friendly! We really should give the user another chance. Luckily, Python lets us convert the entered string into an int <u>tentatively</u> and "catch" the exception if it occurs. For example:

```
try:
    n_max = int(input('Enter a positive integer: '))
except ValueError:
    print('Invalid input')
```

We will put this code inside a while loop to give users as many tries as they need:

```
n_max = -1
while n_max <= 0:
    try:
        n_max = int(input('Enter a positive integer: '))
    except ValueError:
        pass # do nothing
    if n_max <= 0:
        print('Invalid input')
        print('Your input must be a positive integer')
```

We have initially set n_max to -1 so that the program runs through the while loop at least once. This code will reject not only non-integers but also 0 and negative integers.

You can find the complete text of this program in the Sums1toN.py file in the StudentFiles folder.

As you know, Python has the built-in function `sum`. To calculate $1+2+...+n$ we could simply write `sum(range(1, n+1))`. But it would be inefficient to use this call within a loop for n (the same problem as with nested loops). It would also spoil all the fun!

Section 4.3 ~ Exercises

1. Write a program that prints

```
Five little monkeys jumping on the bed,
One fell off and bumped his head.
Mother called the doctor and the doctor said,
"No more monkeys jumping on the bed!"
...

...
One little monkey jumping on the bed,
He fell off and bumped his head.
Mother called the doctor and the doctor said,
"Put those monkeys straight to bed!"
```

⟨ Hints:

1. Use

```
for first_word in ['Five', 'Four', 'Three', 'Two']:
    ...
```

2.

```
print('"No more..."')
```

prints

```
"No more..."
```

⟩

2. Implement the algorithm from Example 1 in Section 4.2 in a Python function. Make your function take one argument, n. Call the function for n = 1000, 10000, and 100000 and compare each returned value times 4 to π.
    ⟨ Hint: `from math import pi` ⟩ ✓

3.   Modify the Sums1toN.py program so that it prompts the user to enter an odd positive integer n_max and prints the sums of positive odd integers $1+3+...+n$ for $n = 1, 3, ...,$ n_max. (Assume that the user always enters a valid number.) For example, if the user enters 9, the program should display

```
1 1
3 4
5 9
7 16
9 25
```

⧼ Hint: for n in range(1, n_max+1, 2): ... ⧽ ✓

4.   Write a program that prompts the user for a positive integer n and prints all positive multiples of 6 (6, 12, 18, etc.) that do not exceed n.

5.   Using Sums1toN.py as a prototype, write a program that prompts the user to enter a positive integer n_max and displays n and $n!$ (*n-factorial*) for n from 1 to n_max. $n! = 1 \cdot 2 \cdot ... \cdot n$. Avoid nested loops and do not use the factorial function from the math module.

6.■   Write a program that prints n, $s_1(n) = 1 + 2 + ... + n$, $s_2(n) = 1^2 + 2^2 + ... + n^2$, and $\dfrac{3s_2(n)}{s_1(n)}$ for $n = 1, 2, 3, ..., 20$. Try to guess the general formula for $1^2 + 2^2 + ... + n^2$ from the resulting table. ✓

7.■   Write a function print_square(n) that displays a "square" with a given side length, made of hyphens and vertical bars (and spaces at the four corners). For example, for $n = 4$, the output should be:

```
 --
|  |
|  |
 --
```

This function does not need a return statement. Use only one for or while loop. (Actually, in Python, you can write this function without any loops, on one line. Can you figure out how?) ✓

8.   Write a function my_pow(x, n) that returns x^n, where $x > 0$ and n is a non-negative integer. Do not use the ** operator or the math.pow function — use one for or while loop. ⧼ Hint: $x^0 = 1$. ⧽ ✓

9.■   Write a function `smallest_divisor` that takes an integer $n > 1$ and returns its smallest divisor that is greater than 1. For example: `smallest_divisor(15)` should return 3; `smallest_divisor(7)` should return 7. Notice that the smallest divisor of *n* is always a prime. ⋛ Hint: `d` is <u>not</u> a divisor of `n` if and only if `n % d is not 0.` ⋚

10.◆   A positive integer is called a *perfect number* if it is equal to the sum of all of its divisors, including 1 but excluding the number itself. For example, $6 = 1 + 2 + 3$. Write a function `sum_of_divisors` that takes a positive integer *n* and returns the sum of all its divisors (excluding *n*).

⋛ Hint:

```
if n % d is 0:
    sum_divs += d
```
⋛

The smallest perfect number is 6. Write a program that finds and prints out the next perfect number.

4.4 Review

Terms introduced in this chapter:

Algorithm	*Nested loops*
Iteration	*Relational operators*
Loop	*Logical operators*

Some of the Python features introduced in this chapter:

```
while <condition>:
for ... in range(last)    # last not included
for ... in range(first, last)
for ... in range(first, last, step)
==, !=, is, is not, <, <=, >, >=, in
and, or, not
```

"""Chapter 5."""

Strings, Lists, Dictionaries, and Files

5.1 Prologue

In programming languages, a *string* represents a word, a line of text, or any combination of characters. Strings are widely used in processing text documents and text files, as prompts to the user and error messages reported by a program, and as elements in data records (such as a student's name, address, and so on).

A *list* represents... well, a list of items. A list in Python can hold numbers, words, strings, or any other objects (even other lists); objects of different types can be intermixed in the same list.

Python treats strings and lists in a unified manner: as iterable "sequences." After all, a string is a sequence of characters and a list is a sequence of its elements. A "sequence" (more precisely an *iterable* object) is a structure in which elements are numbered by integers, called indices, and from which we can request its elements in order: first, second, third, and so on. Python has a special `for ... in ...` loop that allows us to process all the elements of a "sequence" in order. For example:

```
>>> lst = [4, 7, 0]
>>> for e in lst:
        print(e)
4
7
0
>>> s = 'Ted'
>>> for ch in s:
        print(ch)
T
e
d
```

In this chapter we will show you how to access individual characters and "slices" in a string and elements and "slices" in a list.

The last section in this chapter explains how to read and create text files. It is included in this chapter because a text file is a sequence of lines. You can safely skip the section on files until you decide to work on a project that involves files.

5.2 Indices, Slices, and the `in` Operator

The characters in a string and the elements of a list are numbered by integers. A character's or element's number is called its *index*. If s is a "sequence," `s[i]` refers to the element with the index *i*. In Python (as well as in C, C++, Java, and most other programming languages) the numbering starts from 0.

> **If `s` is a sequence (string or list), then `s[0]` refers to the first element of the sequence.**

In Python you can also use negative indices — then you start counting from the end: `s[-1]` is the last element, `s[-2]` is next to last, and so on (see Figure 5-1).

'Green'	'Eggs'	'and'	'Ham'
s[0]	s[1]	s[2]	s[3]
s[-4]	s[-3]	s[-2]	s[-1]

Figure 5-1.   Elements in a list can be indexed by positive and negative numbers

> **The built-in function `len(s)` returns the length of `s`, that is, the number of characters in the string or the number of elements in the list.**

❖   ❖   ❖

Python also allows you to make a "slice" of a sequence — a *substring* of a string or a contiguous chunk of elements of a list. A slice `s[m:n]` is a new sequence made up of the items `s[m], s[m+1], ..., s[n-1]`. Notice that `s[n]` is not included. For example:

```
>>> poem = ['I', 'do', 'not', 'like', 'green', 'eggs', 'and', 'ham']
>>> poem[2:6]
['not', 'like', 'green', 'eggs']
>>> s = 'software'
>>> s[0:4]
'soft'
```

> **The slice `s[m:n]` has n - m elements, provided $0 \leq m < n \leq$ `len(s)`.**

The first or second index in `s[m:n]` can be omitted. Then it takes the default value: 0 for the first index, `len(s)` for the second. For example:

```
>>> s = 'software'
>>> s[:4]   # same as s[0:4]
'soft'
>>> s[4:]   # same as s[4:len(s)]
'ware'
```

> `s[:]` **creates a copy of s.**

You can add a third parameter to a slice: "step." For example, `s[::3]` will take every third element:

```
>>> poem = ['I', 'do', 'not', 'like', 'green', 'eggs', 'and', 'ham']
>>> poem[::3]
['I', 'like', 'and']
>>> s = 'software'
>>> s[1::2]
'otae'
```

A negative step means traverse the sequence backwards. In particular,

> `s[::-1]` **reverses s.**

For example,

```
>>> s = 'software'
>>> rev_s = s[::-1]
>>> rev_s
'erawtfos'
```

❖ ❖ ❖

`in` is a *logical operator* that works with iterable objects. (Recall that the keyword `in` also works with `for`.)

> **x** `in` **s gives** `True` **if x is a character or substring in the string s, or if x is an element of the list s. x** `not in` **s means x is not in s.**

(You can not use `in` to determine whether a "sublist" is part of a list.)

Section 5.2 ~ Exercises

1.   Fill in the blanks.

```
>>> s = 'abcde'
>>> s[2]
```

____

```
>>> s[1:3]
```

____

```
>>> s[1:4:2]
```

____

```
>>> s[:]
```

____

```
>>> s[::-1]
```

____

```
>>> 'a' in s
```

____

```
>>> 'bc' not in s
```

____

2.   Write a one-line function `is_palindrome(word)` that determines whether `word` is a palindrome, that is, reads the same forward and backward. For example, `is_palindrome('racecar')` should return `True`, and `is_palindrome('toyota')` should return `False`. Assume that `word` is in all lowercase letters. Do not use any loops or function calls, just slices. ⸨ Hint:

```
return x is y  # or return x == y
```

returns `True` if x is equal to y; otherwise it returns `False`. ⸩ ✓

3.   Write a one-line function `is_double(s)` that determines whether `s` is a
string that consists of a substring repeated twice. For example,
`is_double('La-La-')` should return `True`, and `is_double('La-La')`
should return `False`. Do not use any loops or function calls (except `len`),
just slices.

4.   What is the value of `len(h[::-1])` if h is `'Happy'`?

(A)   `-1`
(B)   `0`
(C)   `4`
(D)   `5`

5.   Fill in the blanks.

```
>>> 'an' in 'banana'


_______________

>>> 'an' in 'banana'[2:4]


_______________

>>> 'an' in 'banana'[4:0:-1]


_______________

>>> 'aaa' in 'banana'[1::2]


_______________
```

6.   What is the result of

```
>>> [3, 5] in [1, 1, 2, 3, 5, 8]
```

(A)   `3`
(B)   `True`
(C)   `False`
(D)   `SyntaxError`

5.3 Strings

Python uses Unicode to represent strings. A Unicode string represents each character in two bytes. It can encode up to 65,000 characters, enough to encode the alphabets of most world languages and many special characters. For example, *hex* 20ac represents the euro currency symbol '€'. You can enter this symbol in a string as '\u20ac'. (We will discuss the *hexadecimal number system* in Chapter 6.) Try:

```
>>> print('Pecorino di Pienza \u20ac', str(3.95), "l'etto")
Pecorino di Pienza €3.95 l'etto
```

Unicode has a provision to extend the character set even further, into millions of different codes.

> **In Python a character is treated as a string that consists of one character. So, technically speaking, s[i] is a substring of s, same as s[i:i+1].**

A *literal string* is a sequence of characters in quotes. In Python you can use single quotes, as in 'Welcome', double quotes, as in "Welcome", or triple quotes, as in '''Welcome''' or """Welcome""". A literal string within triple quotes can span multiple lines; such strings are often used as *docstrings* in functions (as explained in Chapter 2). We will mostly use single quotes for literal strings, as is customary in Python, unless the string itself contains a single quote, as in "It's Mike's turn". We will use triple double quotes for docstrings.

Literal strings can contain special characters, called *escape characters*. An escape character is represented by a backslash followed by a symbol. For example:

\n	newline	\'	single quote
\t	tab	\"	double quote
\\	backslash		

❖   ❖   ❖

In Python, many functions that help you manipulate strings are implemented in the *object-oriented programming* (*OOP*) manner. In OOP, functions are called *methods*, and they are viewed as being attached to individual objects.

The syntax for calling methods is different: instead of writing `somefun(s)` we write `s.somefun()`. `somefun(s, x)` becomes `s.somefun(x)`. This emphasizes the fact that the first argument is special: it is the object whose method is called. For example:

```
>>> s = 'ooooxxxx'
>>> s.upper()
'OOOOXXXX'
```

`s.find(sub)` returns the index of the first occurrence of the substring `sub` in `s`. `s.find(sub, start)` returns the first occurrence after the index `start`. For example:

```
>>> s = 'One fish two fish'
>>> s.find('fish')
4
>>> s.find('fish', 7)
13
```

`len`, however, is not a method — it is a built-in function. We have to write `len(s)`.

Appendix C summarizes the more commonly used string methods. There are methods that verify whether the string consists of characters from a given category (all letters, all digits, all alphanumeric characters, all white space, and so on), methods that convert the string to upper and lower case, methods that justify the string on the left, right, or in the center of a wider string, methods that find a given substring (or character) in the string, and so on.

Example 1

Write a function that takes a string in the form `'x + y'` and converts it into *postfix notation* `'x y +'`, where x and y are any names, possibly surrounded by spaces or tabs. For example, `to_postfix('  num   +  incr\n')` should return `'num incr +'`.

Solution

```
def to_postfix(s):
    i = s.find('+')
    return s[0:i].strip() + ' ' + s[i+1:].strip() + ' +'
```

`i = s.find('+')` sets `i` to the index of the first occurrence of `'+'` in `s`. `s[i+1:]` returns the tail of the string starting at `s[i+1]` — the same as `s[i+1:len(s)]`. The `strip()` method removes the white space at the beginning and the end of the string. Here `strip` is called for substrings of `s`: `s[0:i]` and `s[i+1:]`

Example 2

Write a function that takes a date in the format m/d/yyyy and returns it in the format dd-mm-yyyy. (In the input string, *m* and *d* represent one- or two-digit numbers for month and day, respectively; in the output string, *mm* and *dd* are two-digit numbers.) For example, `convert_date('7/17/2019')` should return `'17-07-2019'`.

Solution

```
def convert_date(date):
    i1 = date.find('/')
    i2 = date.find('/', i1+1) # or i2 = date.rfind('/')
    return date[i1+1:i2].zfill(2) + '-' + date[0:i1].zfill(2) +  \
                '-' + date[i2+1:]
```

`s.find(sub, startindex)` searches for `sub` starting at index `startindex`; `s.rfind(sub)` looks for the <u>last</u> occurrence of `sub` in `s`. `s.zfill(n)` pads `s` with zeros on the left to make a string of length `n`.

Example 3

Write a function that removes dashes from a string that represents a telephone number. For example, `remove_dashes('800-235-7216')` should return `'8002357216'`.

Solution

```
def remove_dashes(phone):
    return phone.replace('-', '')
```

`s.replace(old, new)` replaces every occurrence of `old` in `s` with `new`. In this case, `new` is an *empty string*.

Example 4

Write a function that replaces every occurrence of `'<b>'` with `'<i>'` and every occurrence of `'</b>'` with `'</i>'` in a given string and returns a new string. For example, `bold_to_italic('Strings are <b>immutable</b> objects')` should return `'Strings are <i>immutable</i> objects'`.

Solution

```
def bold_to_italic(text):
    return text.replace('<b>', '<i>').replace('</b>', '</i>')
```

Don't be surprised by the two dot operators chained in one statement: since `text.replace(...)` returns a string, we can call that string's methods.

Strings are immutable objects. This means none of a string's methods can change its contents.

Consider, for example, the method `upper`, which converts all letters in the string into the upper case. At first glance, it may seem that `s.upper()` changes `s`. But it does not:

```
>>> s = 'amy'
>>> s.upper()
'AMY'
>>> s
'amy'
```

What's going on? `s.upper()` creates a new string `'AMY'` and returns a reference to it, but s remains unchanged. Now try this:

```
>>> s = 'amy'
>>> s2 = s.upper()
>>> s
'amy'
>>> s2
'AMY'
```

To change s to upper case, you need:

```
>>> s = s.upper()
```

Now s refers to the new string in all upper case. The old string is recycled by Python's *garbage collector*.

Likewise,

```
>>> s = 'Hello'
>>> s += ', World!'
```

does not modify s but creates a new string and assigns it to s. Try this:

```
>>> s1 = 'Hello'
>>> s2 = s1   # s2 refers to the same string as s1
>>> s1
'Hello'
>>> s2
'Hello'
>>> s1 += ', World!'
>>> s1   # s1 is now a new string
'Hello, World!'
>>> s2   # still the old string
'Hello'
```

You <u>cannot</u> assign a new value to a character in a string.

For example:

```
>>> s = 'abcd'
>>> s[2] = 'z'
...
TypeError: 'str' object does not support item assignment
```

Section 5.3 ~ Exercises

1. In Sweden, dates are usually written in the "yyyy-mm-dd" format. For instance, 2020-09-01 for September 1, 2020. Write and test a function that converts a date from `'m/d/yyyy'` to `'yyyy-mm-dd'`. ✓

2. Write and test a function that corrects the "two caps" typo: if a string starts with two capital letters followed by a lowercase letter, the function should change the second letter to lower case and return the new string; otherwise it should return the original string unchanged. For example, `fix_two_caps('NEver')` should return `'Never'`.

3. Python strings have the methods `s.startswith(sub)` and `s.endswith(sub)`, which return `True` if `s` starts with `sub` or ends with `sub`, respectively. Pretend that these methods do not exist and write your own functions `startswith(s, sub)` and `endswith(s, sub)`. Do not use loops in your functions: use only `len` and slices:

    ```
    def startswith(s, sub):
        n = len(sub)

        return ________________________
    ```

    Test your functions and make sure they work the same way as the `startswith` and `endswith` methods, including the cases when `sub` is an empty string, when `sub` is the same as `s`, and when `sub` is longer than `s`. ✓

4. Write and test a function `get_digits(s)` that returns a string of all the digits found in `s` (in the same order). For example, `get_digits('**1.23a-42')` should return `'12342'`, and `get_digits('LOL')` should return `' '`, the empty string. ✓

5. Write and test a function that takes a non-empty string of digits and returns a new string with the leading zeros removed. But never remove the last digit.

6. Recall that a valid name in Python can include only letters, digits, and underscore characters and cannot start with a digit. Write a function `is_valid_name(s)` that returns `True` if s represents a valid name in Python; otherwise your function should return `False`. (Do not use Python's library functions.) Test your function on the following strings:

Valid:	Invalid:
`'bDay'`	`'1a'`
`'A0'`	`'#A'`
`'_a_1'`	`' ABC'`
`'_1amt'`	`''`
`'___'`	`'A#'`
`'_'`	`'A-2'`

7. Write a function `is_valid_amt(s)` that returns `True` if the string s represents a valid dollar amount. Otherwise the function should return `False`. A valid amount string is allowed to have white space at the beginning and at the end; the remaining characters must be digits, with the possible exception of one decimal point. If a decimal point is present, it must have at least one digit before it and it must be followed by exactly two digits. For example:

Valid:	Invalid:
`'123.45'`	`'$123.45'`
`'   123.45   '`	`'123.'`
`'   123'`	`'1.23.'`
`'123'`	`'   123.0'`
`'    0.45'`	`'.5'`
`'0'`	`'+0.45'`

Be sure to test your function with all of the above. ✓

8. Write and test a function `remove_tag(s)` that tries to find within s the first substring enclosed within angle brackets, removes it (including the angle brackets) if found, and returns a new string. For example, `remove_tag('Do <u>not</u> disturb')` should return `'Do not</u> disturb'`. If no angle brackets are found, the function should return the original string unchanged. ✓

9. Write and test a function `remove_all_tags(s)` that finds and removes from s all substrings enclosed within angle brackets. ⸤ Hint: Call `remove_tag` from Question 8. ⸥

10.■ A postal code in Canada has the format "ADA DAD" where A is any letter and D is any digit. For example: L3P 7P5. It may have one or several spaces between the first and the second triplet. Write a function that verifies whether a given string has the right format for a valid Canadian postal code, and if so, returns it with all spaces removed and all the letters converted to upper case. If the format is not valid, the function should return `None`. Create a comprehensive set of test strings and test your function on that set.

5.4   Lists and Tuples

In Python a list is written within square brackets, with elements separated by commas. A list can have elements of different types mixed together. For example:

```
[3, 'Large fries', 2.29]
```

The indices and slices and the `in` operator and `len` work for lists the same way as for strings.

> `lst[:]` **returns a copy of** `lst`. **It is the same as** `lst[0:len(lst)]` **or**
> `lst[0:]` **or** `lst[:len(lst)]`.

The built-in functions `min(lst)` and `max(lst)` return the smallest and the largest element in the list `lst`, respectively. For `min` and `max` to work, all the elements of `lst` must be comparable to each other. What is smaller and what is larger depends on the type of objects. Strings, for example, are ordered alphabetically, but all uppercase letters are "smaller" than all lowercase letters.

> **Do not use** `max` **and** `min` **as names for your variables!**

The + operator applied to two lists concatenates them. For example:

```
>>> [1, 2, 3] + [4, 5]
[1, 2, 3, 4, 5]
```

But you can't add one element to a list using +:

```
>>> lst = [1, 2, 3] + 4
...
TypeError: can only concatenate list (not "int") to list
```

You need

```
>>> lst = [1, 2, 3] + [4]
```

or

```
>>> lst.append(4)
```

❖   ❖   ❖

Python also has a built-in function `list(s)` that converts a "sequence" (such as a string) `s` into a list and returns that list. For example:

```
>>> list('123')
['1', '2', '3']
```

> **Do not use the name `list` for your variables!**

❖   ❖   ❖

> **Lists are <u>not</u> immutable: a list has methods that can change it.**

A list has methods that append an element to it, remove an element from it, find a given value, reverse the list, and sort the list (arrange its elements in order). For example:

```
>>> lst = [1, 3, 2]
>>> lst
[1, 3, 2]
>>> lst.append(7)
>>> lst
[1, 3, 2, 7]
>>> lst.reverse()
>>> lst
[7, 2, 3, 1]
>>> lst.sort()
>>> lst
[1, 2, 3, 7]
>>> lst.sort(reverse=True)
>>> lst
[7, 3, 2, 1]
```

You can also assign a new value to a list element. For example:

```
>>> lst = [1, 4, 2]
>>> lst[1] = 3
>>> lst
[1, 3, 2]
```

Appendix D summarizes commonly used list methods.

The `lst.index(x)` method returns the index of the first element in `lst` that is equal to `x`. When `x` is not in `lst`, this method raises an exception. So it may be necessary to check first whether `x` is in `lst`:

```
if x in lst:
    i = lst.index(x)
else:
    ...
```

A string also has a method called `index`. Like `lst.index(x)`, string's `index(x)` raises an exception when the target is not found. So it is usually better to use the `find` method, which returns –1 when the target is not found. Lists do not have a method `find`.

> **The statement `del lst[i]` removes `lst[i]` from `lst`. You can also remove a slice: `del lst[i:j]`.**

For example:

```
>>> poem = ['I', 'do', 'not', 'like', 'green', 'eggs', 'and', 'ham']
>>> del poem[1:3]
>>> poem
['I', 'like', 'green', 'eggs', 'and', 'ham']
```

Example 1

Write a function that swaps the first and last elements of a given list.

Solution

```
def swap_first_last(lst):
    lst[0], lst[-1] = lst[-1], lst[0]
```

Example 2

Write a function that rotates a list, moving its first element to the end.

Solution

```
def rotate_left(lst):        Or:        def rotate_left(lst):
    n = len(lst)                            n = len(lst)
    temp = lst[0]                           temp = lst[0]
    for i in range(1, n):                   lst[0:n-1] = lst[1:n]
        lst[i-1] = lst[i]                   lst[n-1] = temp
    lst[n-1] = temp
```

Or:

```
def rotate_left(lst):
    lst.append(lst.pop(0))
```

❖   ❖   ❖

We often need to apply the same formula or function to each element of a list and create a new list that contains these new values. Python has a feature known as *list comprehensions*, which provides shorthand for such procedures. For example:

```
result = [2*x for x in lst]
```

This is the same as:

```
result = []
for x in lst:
    result.append(2*x)
```

Another example:

```
lengths = [len(w) for w in words]
```

If `words` is `['All', 'is', 'well']`, then `lengths` is `[3, 2, 4]`.

This is the same as:

```
lengths = []
for w in words:
    lengths.append(len(w))
```

You can also apply a condition to the elements and include in the resulting list only the elements for which the condition is true. For example:

```
>>> nums = [-1, -4, 2, 5, -3, 11]
>>> positives = [x for x in nums if x > 0]
>>> positives
[2, 5, 11]
```

Example 3

Write a function that takes a list of words and creates a new list of words choosing only those words whose length is not less than 3 and not greater than 5.

Solution

```
def choose_length_3_5(words):
    return [w for w in words if 3 <= len(w) <= 5]
```

(Notice that in Python you can combine two relational operators in one expression. This feature is not common in other programming languages. We will discuss relational operators and `if` statements in Chapter 7.)

As we said earlier, lists are not immutable: a list has methods that add and remove elements, and so on. A *tuple* is an object that represents an *immutable* list. A tuple is written by placing its elements in parentheses, separating them by commas. For example: `(3, 4, 5)`. To create a tuple with only one element, follow the value with a comma, to distinguish the tuple from a number in parentheses. For example:

```
>>> t = (0,)
```

Indices, slices, +, the `in` and `not in` operators, and `len` work for tuples the same way as for lists, but tuples have no methods except `count` and `index`. Tuples are convenient when you need to create a set of lists, because sets can only hold immutable objects.

Example 4

Write a function that takes a list of words and returns a tuple containing the shortest and the longest word.

Solution

```
def shortest_and_longest(words):
    shortest = 100*'x'  # longer than any reasonable word
    longest = ''  # empty string
    for w in words:
        if len(w) < len(shortest):
            shortest = w
        if len(w) > len(longest):
            longest = w
    return shortest, longest
```

Now

```
print(shortest_and_longest(['You', 'are', 'my', 'sunshine']))
```

displays

```
('my', 'sunshine')
```

> **Python packs multiple values into a tuple automatically and unpacks a tuple automatically when it is assigned to multiple values.**

That's why you could write

```
    return shortest, longest
```

and

```
word1, word2 = shortest_and_longest(['You', 'are', 'my', 'sunshine'])
```

— it will set `word1` to `'my'` and `word2` to `'sunshine'`.

❖   ❖   ❖

We say that a list is *sorted* if its elements are arranged in order of values.

Sorting is a very common operation. For example, if you have two mailing lists and want to merge them into one and eliminate duplicate records, it is better to sort each list first. There are several sorting algorithms, ranging from the very straightforward (see Question 6) to the more advanced. Python provides the built-in function `sorted` that, when applied to an iterable sequence `s`, returns a <u>list</u> with all the elements from `s` arranged in ascending (increasing) order. A list also has the method `sort` that sorts the list (in place) in ascending order. Both `sorted` and `sort` will arrange the elements in descending (decreasing) order when given the optional parameter `reverse=True`. For example:

```
>>> coins = [1, 10, 5, 25]
>>> coins.sort(reverse=True)
>>> coins
[25, 10, 5, 1]
```

Section 5.4 ~ Exercises

1. Pretend that list's `reverse` method and the built-in function `reversed` do not exist and write your own function `reverse(lst)` that reverses the list in place and returns `None`. ⸮ Hints: proceed from both ends, but not too far... No need for a `return` statement. ⸮ ✓

2. Write and test a function that returns the index of the largest element in the list (or, if several elements have the largest value, the index of the first one of them). Do not use the built-in function `max`. ⸮ Hint: see Example 4. ⸮

3.■ Write and test a function that takes a list with three or more elements and returns the index of the element that makes the greatest sum with its left and right neighbors.

4. Replace the following code with a list comprehension:

```
newnums = []
for x in nums:
    newnums.append(x + 1)
```

✓

5.   Write and test a function `all_pairs(n)` that returns the list of all $\dfrac{n(n-1)}{2}$ tuples `(i, j)`, where $1 \leq i < j \leq n$. For example, `all_pairs(3)` should return `[(1,2), (1,3), (2,3)]`.

6.   Pretend that list's `sort` method does not exist and write your own function `sort(a)` that sorts list `a` in ascending order (and returns `None`). Use the following algorithm, which is called *Selection Sort*:

> Set *n* equal to `len(a)`
> Repeat the following steps while *n* > 1:
>   Find the largest among the first *n* elements (see Question 2)
>   Swap it with the *n*-th element (`a[n-1]`)
>   Subtract 1 from *n*

7.   The function below takes sorted lists `a` and `b` and merges all the elements from `b` into `a`, so that the resulting list is sorted:

```
def merge(a, b):
    for x in b:
        i = 0
        while i < len(a) and x > a[i]:
            i += 1
        a.insert(i, x)
```

If `a = [1, 3, 5, ..., 199]` and `b = [2, 4, 6, ..., 200]`, how many times will the comparison `x > a[i]` be executed in `merge(a, b)`? Make a couple of minor changes to the above code to improve its efficiency: `x > a[i]` should be executed no more than 200 times for these lists. ⸤ Hint: move one line and add one line of code. ⸥ ✓

5.5   Dictionaries

A *dictionary* in Python establishes a correspondence between a set of *keys* and a set of *values*. Only one value corresponds to each key. In mathematical terms, a dictionary is like a function on a set of keys. In practical terms, a dictionary lets you quickly look up a value (an object, a data record, a text segment) associated with a given key. In a database of taxpayers, for example, the key may be a taxpayer's social security number, and his or her record may be the associated value. In the zip code lookup program, the key may be a zip code, and the name of the city or town for that zip code will be the associated value.

The set of keys is implemented as a *hash table* — a data structure optimized for efficient searching. You have probably heard of hashtags on Twitter.

To create a dictionary with initial entries, list the *key* : *value* pairs within curly braces. For example:

```
coins = {'Quarter': 25, 'Dime': 10, 'Nickel': 5, 'Penny': 1}
```

Another example:

```
cubes = {1: 1, 2: 8, 3: 27, 4: 64, 5: 125}
```

Once a dictionary is defined, you can refer to its values using a key as an "index." For example:

```
>>> coins['Dime']
10
```

(Notice that `'Dime'` here is a literal string, not a name of the key.)

> **Use the syntax with brackets, just like you do with lists, to get or modify the value associated with a key. If `k` is not in `d`, `d[k]` = `x` will add the `(k, x)` pair to `d`.**

`d = {}` makes `d` an empty dictionary.

Think of dictionary keys as a `set` (although, strictly speaking, its type is `dict_keys`).

> **`len(d)` returns the number of key-value pairs in `d`. `k in d` is `True` if `k` is a valid key in `d`. `del d[k]` deletes `k` and the value associated with it from `d`.**

> **Notice that a value associated with a key may be a list, a tuple, a set — any object.**

For example:

```python
spanish_english = {
    'abajo' : ['down', 'below', 'downstairs'],
    'presto' : ['quick', 'prompt', 'ready', 'soon']
}

print(spanish_english['presto'])
```

The output will be:

```
[quick, prompt, ready, soon]
```

Example 1

Write a function `count_letters(text)` that counts the number of occurrences of each letter in `text` and returns a dictionary of the counts. For example, `count_letters('MISSISSIPPI')` should return a dictionary equal to

```
{'M': 1, 'I': 4, 'S': 4, 'P': 2}
```

Solution

```python
def count_letters(text):
    """Return the dictionary of pairs (letter, count) where
       count is the number of occurrences of letter in text.
    """
    counts = {}  # start with an empty dictionary

    for letter in text:
        if letter in counts:
            counts[letter] += 1
        else:
            counts[letter] = 1

    return counts
```

A dictionary has the convenient method `get`: `d.get(k, default_value)` returns `d[k]` if `k` is in `d`; otherwise it returns `default_value`. The `for` loop in the above code can be shortened using `get`:

```python
    ...
    for letter in text:
        counts[letter] = counts.get(letter, 0) + 1
    ...
```

> `set(d)` and `list(d)` return the set and the list, respectively, of <u>keys</u> in d.
>
> `set(d.values())` and `list(d.values())` return the set and the list, respectively, of all values in d.
>
> `set(d.items())` and `list(d.items())` return the set and the list, respectively, of (key, value) pairs in d.

Common dictionary methods are summarized in Appendix D. Enter `help(dict)` or consult Python's documentation to get a complete list of dictionary methods.

Section 5.5 ~ Exercises

1. Define a dictionary that holds several name-password pairs (with the name used as the key). Write and test a function that takes such a dictionary and a (name, password) pair (represented as a tuple) and returns `True` if the name is in the dictionary and the password matches the one in the dictionary.

2. A song is described by its title, band, and duration in seconds. Define a dictionary for "Alligator" by The Nationals, 4:05. ✓

3. Suppose in a dictionary all values associated with keys are different. Write a function `reverse_dictionary(d)` that takes such a dictionary and returns a "reverse" dictionary in which each value becomes a key and its key becomes the associated value. For example,
    `reverse_dictionary({1:'A', 2:'B'})` should return
    `{'A':1, 'B':2}`. ✓

4.■ Define a dictionary that can serve as an index for a book. Each key is a word; the associated value is a list of the page numbers of all the pages on which that word occurs. Write and test a function
    `add_entry(d, word, page)` that takes such a dictionary d and adds `page` to the list of page numbers associated with `word`. If `word` is already in the dictionary and `page` is in its list, then `page` should not be added. If `word` is not yet in the dictionary, then `add_entry` should create a new entry for `word` with `page` in its list.

5.6 Files

A *file* is a collection of related data stored on a computer's hard disk, a memory stick, another digital device, computer-readable media, or a remote server on the Internet. A file can store a text document, a song, an image, a video clip, the source code for a program, and so on. Computer programs can read files and create new files.

A file is identified by its name and extension. The extension often identifies the format and/or purpose of the file. For example, in `Birthdays.py`, the extension `py` tells us that this file holds the source code for a Python program, and in `graduation.jpg`, the extension `.jpg` tells us that this file holds an image in the JPEG format.

In general, when a program creates a file, it is up to that program to decide how the data will be organized in the file. But there are some standard file formats that many different programs understand. You have probably encountered `.mp3` files for music, `.jpg` files for images, `.docx` files for *Word* documents, `.htm` or `.html` files for web pages, and so on.

All files fall into two broad categories: *text files* and *binary files*. A text file holds lines of text, separated by end-of-line markers. The characters in a text file are encoded in one of the standard encodings, such as ASCII (read 'as-kee) or one of the Unicode encodings. An end-of-line marker is usually a newline character `'\n'`, a carriage return character `'\r'`, or their combination, `'\r\n'`. Files with the extensions `.txt`, `.html`, and `.py` are examples of text files.

A binary file holds any kind of data, basically a sequence of bytes — there is no assumption that the file holds lines of text (although you can treat a text file as a binary file, if you wish). `.mp3`, `.jpg`, and `.docx` files are examples of binary files.

An operating system organizes files into a system of nested folders (also called *directories*) and allows users to move, copy, rename, and delete files. Each operating system also provides services to programs for reading and writing files. Programming languages usually include libraries of functions for dealing with files. Here we will use only Python's built-in functions for reading and writing text files.

To open a text file in a Python program, use the built-in function `open`. For example:

```
f = open('GreenEggs.txt')
```

> **The file must be in the folder that the Python interpreter uses as its current directory. See Appendix A at `www.skylit.com/python` for instructions on how to choose that folder.**

Alternatively you can specify the *absolute pathname* of the file. For example:

```
f = open('C:/PythonProjects/GreenEggs.txt')
```

This is not recommended, however, because your program will stop working if the file is moved to a different folder or the folder is renamed.

> **When you have finished working with the file, you have to close it by calling `f.close()`.**

❖　❖　❖

When a text file is open for reading, Python treats it as a sequence of lines. To read the next line, use file's `readline` method. Try, for example:

```
>>> f = open('GreenEggs.txt')
>>> line = f.readline()
>>> line
'I am Sam\n'
```

> **Notice that Python leaves the newline character `'\n'` in the string read from a file. If there are no more lines left in the file, the next call to `readline` returns an empty string.**

The easiest way to read and process all the lines from a text file is with a `for` loop:

```
for line in f:
    ...  # process line
```

Example 1

```
for line in f:
    print(line, end='')
```

The `end=''` parameter tells `print` not to add a newline character to the output. We want it here because every line read from a file already has a newline character at the end. Without `end=''` the file would be printed out double-spaced.

Example 2

A file object also has the method `readlines`, which reads all the lines from the file and returns a list of all the lines read.

```
>>> f = open('GreenEggs.txt')
>>> lines = f.readlines()
>>> f.close()
>>> len(lines)
19
>>> lines[0]
'I am Sam\n'
>>> lines[1]
'Sam I am\n'
>>> lines[-1]
'I do not like them, Sam-I-am.\n'
```

If possible, read a file line by line. Avoid reading the whole file into a list because a file may be quite large.

None of the "read" methods skip empty lines; an empty line is returned as `'\n'`.

To create a file for writing, add a second argument, `'w'`, to the call to `open`. Use file's `write` method to write a string to a file.

Example 3:

```
f = open('Story.txt', 'w')
f.write('I do not like\n')
f.write('green eggs and ham.\n')
f.close()
```

The result is a file called `story.txt` that contains

```
I do not like
green eggs and ham.
```

> **Be careful: if you open for writing a file that already exists, it will be reduced to zero size and the old information in the file will be lost!**

> **Don't forget to close the file when you have finished writing into it. If you forget to close the file, some of the information may not be written into it.**

Instead of `f.write(s)`, you can use `print(s, file=f)`. As you know, `print` adds one newline character at the end of each line (unless a trailing comma is supplied in the `print` statement), so you don't have to supply `'\n'` in the lines you print.

Example 4

```
f = open('Story.txt', 'w')
print('I do not like', file=f)
print('green eggs and ham.', file=f)
f.close()
```

Example 5

```
f = open('Transactions.txt', 'w')
qty = 3
price = 2.95
amt = qty * price
print(' {0:4d}   {1:5.2f} {2:7.2f}'.format(qty, price, amt), file=f)
qty = 4
price = 1.85
amt = qty * price
print(' {0:4d}   {1:5.2f} {2:7.2f}'.format(qty, price, amt), file=f)
f.close()
```

This creates a file `Transactions.txt` that contains

```
3    2.95    8.85
4    1.85    7.40
```

Section 5.6 ~ Exercises

1.   Write a program that reads a text file and displays it with a line number at the beginning of each line. ✓

2.   Write a program that creates a copy of a text file. Your program should prompt the user to enter the names of the input and output files.

3.   Write a program that reads a given file and displays all the lines that contain a given string. ✓

4.■   You have two input files, each sorted in the order established by the `<=` operator: if `line1` is earlier in the file than `line2`, then `line1 <= line2` is true. (This ordering is basically alphabetical, except any uppercase letter is "smaller" than any lowercase letter, and any digit is smaller than any letter.) Write and test a program that merges these files into one sorted file. Do not load the files into memory — read them line by line.

5.▪   Write a program that counts the occurrences of the 26 English letters, case-blind, in a text file and displays their relative frequencies. The program should prompt the user for a file name, count the number of occurrences of each letter, and display each letter, along with its frequency expressed as a percentage of the total number of letters. The letters should be displayed in order of frequency. For example, if the file contains two 'A's and three 'B's, and no other letters, then the output should be:

```
Text file name: letters.txt
B: 60.0
A: 40.0
C:  0.0
...
Z:  0.0
```

⸨ Hints:

1. Allocate a list of 26 counts, initially set to 0: `counts = 26*[0]` or use a dictionary.

2. Read the file one line at a time. For each line read, increment the appropriate count for each letter in the line. The index of the count to be incremented is the index of the letter (converted to the upper case) in `abc = 'ABCDEFGHIJKLMNOPQRSTUVWXYZ'`.

3. The built-in `sum` function returns the sum of all the elements of a list.

4. Create a list of (count, letter) pairs. The built-in function `zip(s1, s2)` combines the respective elements from the sequences `s1` and `s2` into pairs (tuples) and generates a sequence of such tuples. For example, `sorted(zip([40.0, 60.0], ['A', 'B']), reverse=True)` makes `[(60.0, 'B'), (40.0, 'A')]`.

5. Print out both elements of each pair in the desired format.

⸩

6.◆   Write a program that quizzes a student on the capitals of the 50 U.S. states. Create a text file that lists all the states and their capitals. The program reads the file line by line and saves all the state-capital pairs in a dictionary. The program then extracts a set of all the states from the dictionary, and ten times presents a randomly chosen state to the student (without repetition). The program matches each answer (case blind) against the state capital in the dictionary and keeps track of the number of correct answers. At the end, the program displays the number of correct answers.

5.7 Review

Terms introduced in this chapter:

String   *Dictionary*
List   *File*
Tuple   *Text file*
Index   *Binary file*
Slice
List comprehension

Some of the Python features introduced in this chapter:

```
s[i], s[i:j], s[:i], s[i:], s[:]
s.isalpha(), s.isdigit(), s.isalnum(), ...
s.upper(), s.lower(), ...
s.replace(old, new), s.strip()
s.find(sub), s.find(sub, start), s.rfind(sub)

for x in s:
    ... # process x

lst = []
lst_copy = lst[:]
lst.append(x), lst.insert(i, x), lst.pop(), lst.pop(i)
lst.remove(x)
del lst[i], del lst[i:j], lst.reverse()
lst.sort(), sorted(s), sorted(s, reverse=True)
[g(x) for x in lst]
[g(x) for x in lst if ...]

(a, b, c), (a,)

d = {k1: value1, k2: value2, k3: value3}
d[k] = new_value
len(d)
k in d
del d[k]
d.get(k, default_value)
```

```python
f = open('MyInputFile.txt')
f.readline(), f.readlines()

for line in f:
    ... # process line

f = open('MyOutputFile', 'w')
f.write(s)
print(line, file=f)

f.close()
```

```
>>> chapter = 0b110
>>> chapter
6
>>>
```

Number Systems

6.1 Prologue

What is 5, mathematically speaking? There are many possible answers. Five is how many fingers I have on my right hand. Five is what follows four. And five is the property that all sets of "five" objects have in common. But the last definition seems to be circular: How can we tell that two sets have the same number of elements without counting? And to count, you already need to know one, two, three, four, five...

It turns out, however, that there is a simple method of comparing the numbers of elements in two sets without counting. If we can establish a correspondence between the two sets, such that each element of the first set is paired up with exactly one element of the second set and vice versa, then we can be sure the two sets have the same number of elements. Suppose, for example, that there are several students and several backpacks on the playground. Ask each student to pick up one backpack. If there are not enough backpacks for all the students, the set of students has more elements, and if there are some backpacks left on the ground, the set of backpacks has more elements. If every student gets one backpack, and there are no backpacks left, then the number of students and the number of backpacks are the same. Another example: if you can touch with each finger of one hand a different object in a set, and no objects remain untouched, then the set has five elements.

It would be nice to create one special set of five different objects, so that we could compare every other set to it. Suppose the world is populated only by sets, as in pure set theory. How can we devise five different objects in such a world? Take the empty set $\varnothing$ — call it 0. Now take a set of one element: $\{\varnothing\}$ — call it 1. Now take the set that contains 0 and 1 as elements: $\{\varnothing, \{\varnothing\}\}$ — call it 2. $\{\varnothing, \{\varnothing\}, \{\varnothing, \{\varnothing\}\}\}$ is 3. And so on.

6.2 Positional Number Systems

Once we have figured out, more or less, what a number is, the question of how to represent different numbers remains. What is 34, for example? This combination of two digits represents a crucial invention of human civilization, a *positional number system*. Without it we would still be writing 34 as XXXIV or something like that, whereas in our system it is easy to see: 34 is three times ten plus four. Ten is a magic number, the *base* of our positional system. We probably picked ten simply because humans have ten fingers. If you need to tell a stranger who doesn't speak your language that you are willing to swap your iPad for 34 apples, it is convenient to gesture three times with both hands open and then once more with four outstretched fingers. So the importance of ten is a human universal constant across languages and cultures.

If you have a number $\underbrace{d_n...d_2d_1d_0}_{\text{decimal digits}}$, then that number is equal to

$$d_n \cdot 10^n + ... + d_2 \cdot 10^2 + d_1 \cdot 10^1 + d_0 \cdot 10^0.$$ For example, $347 = 3 \cdot 10^2 + 4 \cdot 10 + 7$.

Now imagine what would happen if we humans had only three fingers. We would count 1, 2, ... and then what? Three would become our magic number! After 2 we would write 10, 11, 12, then 20, 21, 22, and then 100. 21_3 (21 base 3) would mean $2 \cdot 3 + 1 = 7_{10}$ (7 base 10). 14 would become 112, and instead of 34 we would write 1021 ($1021_3 = 1 \cdot 3^3 + 0 \cdot 3^2 + 2 \cdot 3 + 1 = 34_{10}$).

The "base 3" system would have its advantages. There would be fewer digits to learn, and it would be easier to learn how to count: one, two, ten, eleven, twelve, twenty, and so on. And you could live to celebrate your 100th birthday (our age 9), 1000th birthday (our age 27), and even 10000th birthday (our age 81). On the other hand, your phone number, instead of 10 digits, would need 20 or 21 — hard to remember.

We can do arithmetic on numbers written in base 3 (or any other base) in the same manner as in base 10.

Example 1

Calculate $2211_3 + 102_3$

Solution

$$
\begin{array}{r}
2211 \\
+102 \\
\hline
10020
\end{array}
$$

$1 + 2 \,(= 3_{10}) = 10_3 \;\Rightarrow\; $ write 0, carry 1.

$1 + 0 + 1 \,(= 2_{10}) = 2_3 \;\Rightarrow\; $ write 2, carry 0.

$2 + 1 + 0 \,(= 3_{10}) = 10_3 \;\Rightarrow\; $ write 0, carry 1.

$2 + 0 + 1 \,(= 3_{10}) = 10_3 \;\Rightarrow\; $ write 0, carry 1.

Write 1.

Example 2

Calculate $2 \cdot 1423_5$

Solution

$$
\begin{array}{r}
1423 \\
\times2 \\
\hline
3401
\end{array}
$$

$3 \cdot 2 \,(= 6_{10}) = 11_5 \;\Rightarrow\; $ write 1, carry 1.

$2 \cdot 2 + 1 \,(= 5_{10}) = 10_5 \;\Rightarrow\; $ write 0, carry 1.

$2 \cdot 4 + 1 \,(= 9_{10}) = 14_5 \;\Rightarrow\; $ write 4, carry 1.

$2 \cdot 1 + 1 \,(= 3_{10}) = 3_5 \;\Rightarrow\; $ write 3.

Section 6.2 ~ Exercises

1. Find all two-digit numbers (base 10) such that the number is equal to twice the sum of its digits. Then find all two-digit numbers (base 10) such that the number is equal to 7 times the sum of its digits. ✓

2. Take any number and subtract from it the sum of all its digits (base 10). Show that the difference is always evenly divisible by 9.

3.■ In a Sudoku puzzle you need to fill a 9-by-9 grid with numbers in such a way that each row, each column, and each of the nine 3-by-3 squares contains all the numbers from 1 to 9. Take any completed Sudoku grid and combine the first three columns into one column of 3-digit numbers. Do the same for the next three columns and the last three columns. Prove that the sum of the numbers in all three new columns is the same. What is the value of that sum? ✓

4. Write 15_{10} and 24_{10} in base 3. ✓

5. Write 121_3 and 2020_3 in base 10.

6. Write 10011100_2 in base 10. ✓

7. Subtract 1 from 10000_3 and write the answer in base 3.

8.■ Write 243_5 in base 7.

9. Calculate $201_3 + 12_3$ and $201_3 - 12_3$ without converting the numbers into decimal (base 10) representation. ✓

10.■ How can you quickly tell whether a number written in base 3 is even (that is, evenly divisible by 2)? ✓

6.3 The Binary, Octal, and Hexadecimal Systems

The next step is to pretend that we don't have any fingers at all, just two hands. Then we would have only two digits, 0 and 1. This base 2 system is called the *binary system*. In the binary system, non-negative integers are represented like this:

Base 10:	Binary:
0	0
1	1
2	10
3	11
4	100
5	101
6	110
7	111
8	1000
. . .	. . .

It is very convenient to use binary representation of numbers in computers, because, as we know, each bit in the computer memory can represent only two values, 0 and 1. However, binary numbers are hard to work with for humans because even small numbers are far too long in binary representation. For example, $98_{10} = 1100010_2$. We need a compromise: a system that is easy to convert to binary and also easy to read for humans (at least for programmers).

There are two such systems: *octal* and *hexadecimal*. The octal system uses base 8 and has eight digits: 0 - 7. Each of these digits can be represented by a three-digit binary number (with leading zeros):

0	000
1	001
2	010
3	011
4	100
5	101
6	110
7	111

To convert an octal number into binary, simply replace each octal digit with its 3-digit binary representation. Experienced programmers know the bit patterns for the octal digits and can translate them instantaneously.

Example 1

    3561₈ = 011 101 110 001₂ = 011101110001₂

❖   ❖   ❖

The reverse conversion, from binary to octal, is also easy: split the string of binary digits into triplets, starting from the rightmost digit (add leading zeros on the left if necessary), then write the corresponding octal digits.

Example 2

    11010111100₂ = 011 010 111 100₂ = 3274₈

❖   ❖   ❖

The hexadecimal system is the base 16 system. Its 16 digits are 0, 1, 2, 3, 4, 5, 6, 7, 8, 9, A, B, C, D, E, and F. A stands for 10, B for 11, and so on; F stands for 15. Each of these digits can be represented by a four-digit binary number, as follows:

```
0        0000
1        0001
2        0010
3        0011
4        0100
5        0101
6        0110
7        0111
8        1000
9        1001
A        1010
B        1011
C        1100
D        1101
E        1110
F        1111
```

To convert a "hex" (hexadecimal) number into binary, replace each hex digit with its 4-digit binary representation. To convert a binary number into hex, split the string of its binary digits into quads (four-digit strings), starting from the right, and replace each quad with its equivalent hex digit. Again, experienced programmers remember the bit patterns for the hex digits and can do such conversions very quickly.

Example 3

```
B6F8₁₆ = 1011 0110 1111 1000₂

0111 0110 1110 1010₂ = 76EA₁₆
```

❖   ❖   ❖

These days, octal numbers are rarely used; the hex system is far more popular, simply because you can conveniently represent the value of one byte (eight bits) as two hex digits.

To convert a binary number into base 10 manually, it is faster to first convert it into hex, then from hex into base 10.

Example 4

$$111010010_2 = \text{hex } 1D2 = 1 \cdot 16^2 + 13 \cdot 16 + 2 = 256 + 13 \cdot 16 + 2 = 466$$

The same is true for converting decimal into binary: first convert the number into hex, then from hex into binary.

❖   ❖   ❖

Programming languages often support binary, octal, and hexadecimal constants. In Python, for example, a sequence of octal digits preceded by `0o` (zero and the letter 'o') represents a number in base 8. Try it:

```
>>> 0o46 # 4 and 6 are octal digits
38
>>> -0o46
-38
```

A hex number is written with the `0x` prefix. You can use uppercase or lowercase letters for the hex digits:

```
>>> 0x1D2 # 1, D, 2 are hexadecimal digits
466
>>> 0x1d2
466
```

In Python you can also use binary constants. A binary constant is written as the `0b` prefix followed by binary digits — for example, `0b100110`.

Python has a way to convert a number into a string of hex, octal, or binary digits. Try:

```
>>> hex(466)
'0x1d2'
>>> '{0:x}'.format(466)
'1d2'
>>> '{0:04x}'.format(466)
'01d2'

>>> oct(38)
'0o46'
>>> print('{0:o}'.format(38))
46

>>> bin(38)
'0b100110'
>>> '{0:08b}'.format(38)
'00100110'
```

You will often see hex numbers written with leading zeros, because hex numbers often represent the contents of a fixed-length group of bytes (two hex digits for each byte).

The built-in function `int` provides a way to convert a string of digits, written in any given base, into an `int` value. Try:

```
>>> int('46')   # Default base 10
46
>>> int('46', 8)
38
>>> int('1d2', 16)
466
>>> int('111010010', 2)
466
```

"Binary" arithmetic is similar to regular arithmetic (Figure 6-1). To add two binary numbers, write them one underneath the other with the unit digits aligned. Add the unit digits. If the result is 2, subtract 2 and set the carry bit. Write the result. Add the 2s digits and the carry bit. If the result is 2 or 3, subtract 2 and set the carry bit. Write the result. And so on. Subtraction is similar.

```
  0 0 0 0 1 0 0 0 0 1 1 1  ⟵        Carry bit
   1011010100011
 +
         11000111
 ─────────────────────
   1011101101010
```

Figure 6-1. Binary addition

Section 6.3 ~ Exercises

1.   What is the largest integer that has 7 binary digits? 15 digits? ✓

2.   Convert hex 90AB into binary.

3.   Convert binary 01011101011 into hex.

4.   Convert hex F0C2 into octal. ✓

5.   What are the results, written in base 2, of $1011100_2 * 4_{10}$ and $1011100_2 / 4_{10}$? ✓

6.▪   Write and test a function that counts and returns the number of bits (binary digits) set to 1 in a positive integer n <u>without converting n to a string</u>.

```python
def count_bits(n):
    """Return the count of bits set to 1 in a positive
        integer.
    """

    ...
```

For example, `count_bits(12)` should return 2, because $12 = 00001100_2$.

⟨ Hint: Check the digits from right to left; `divmod(n, 2)` returns `n//2` and the rightmost binary digit; `n//2` shifts the binary digits of n to the right by one and gets rid of the rightmost digit. ⟩

6.4 Representation of Numbers in Computers

Integers and floating-point numbers are typically represented in a manner convenient for popular microprocessors. But Python 3 is different:

> **In Python 3, large integers are automatically extended to "unlimited" range.**

An integer value typically takes 32 bits (4 bytes) or 64 bits (eight bytes). If we use 32 bits for representing unsigned (non-negative) integers in binary, we are able to represent all the numbers in the range from 0 to $2^{32} - 1$. For signed numbers, positive and negative, the most significant bit typically represents the sign. If we use 32 bits to represent signed integers, then positive values range from 0 to $2^{31} - 1$, with the most significant bit set to 0. The numbers in the range from 2^{31} to $2^{32} - 1$ (that is, the numbers with the most significant bit set to 1) are interpreted as negative. The representation of negative numbers is called *two's complement*: unsigned n is interpreted as $n - 2^{32}$. In the two's complement system, `11....111` (thirty-two 1s) is –1; `11....110` is –2; and so on. The most significant bit becomes the sign (Figure 6-2).

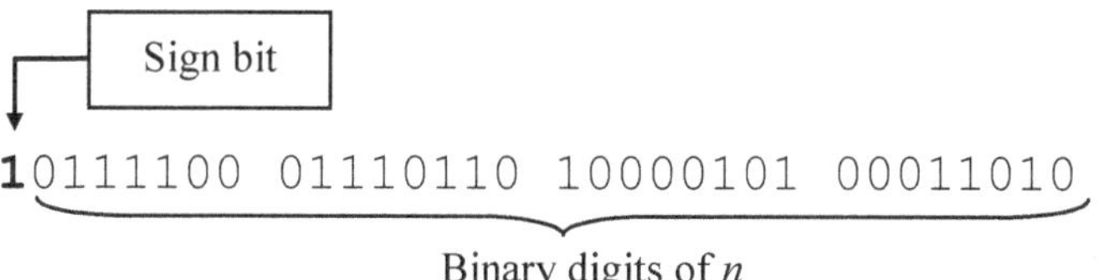

If the sign bit is 0, the number is interpreted as n (non-negative).
If the sign bit is 1, the number is interpreted as $n - 2^{32}$ (negative).

Figure 6-2. Representation of integers in 32 bits. Negative numbers are represented using the two's complement system.

The two's complement system is convenient because $(2^{32} - x) + x = 2^{32}$, which is outside of the 32-bit range; the most significant bit gets lost, and the result is 0. The CPU can perform arithmetic as if operating on 32-bit unsigned numbers — it is not even aware of what we interpret as negative numbers.

A `float` usually takes 64 bits (8 bytes) and represents a number in the standard way expected by the CPU. Like scientific notation, this representation consists of a sign, a fractional part (*mantissa*) and an *exponent* part, but here both the mantissa and the exponent are represented as binary numbers. The IEEE (Institute of Electrical and Electronics Engineers) standard for an 8-byte (64-bit) representation uses 1 bit for the sign, 11 bits for the exponent and 52 bits for the mantissa. 1023 is added to the exponent to ensure that negative exponents are still represented by non-negative numbers.

This 8-byte format allows us to represent numbers in the range from approximately -1.8×10^{308} to 1.8×10^{308}, with about 17 decimal digits of precision. Note that you can represent in this format "only" 2^{64} different numbers, while there are infinitely many real numbers. Conversion from decimal to binary may introduce a small error.

There are many apps on the Internet that illustrate binary and hex representation of numbers and allow you to play with number conversions.

With a little work, we can also represent *rational numbers* in a computer. A rational number is a fraction with an integer numerator and denominator. Python has a module called `fractions` that implements a class of objects `Fraction`. `Fraction` implements arithmetic operations on fractions. For example:

```
>>> from fractions import Fraction
>>> f1 = Fraction(2, 3)   # Creates f1 = 2/3
>>> f2 = Fraction(1/2)    # Creates f2 = 1/2 -- also works
>>> print(f1 + f2)
7/6
>>> print(f1*f2)
1/3
```

Section 6.4 ~ Exercises

1. If integers are represented in four bytes and negative numbers are represented in the two's complement system, what number does $FFFFFACE_{16}$ represent? ✓

2. Explain the result of

    ```
    >>> 1000000000.0 + 0.0000000001
    ```

    ✓

3. Explain the result of

    ```
    >>> 3 * 0.1 == 0.3
    ```

4. Find the smallest positive integer n such that in Python
    `int(17.0 / n * n)` is not `17.0`. ✓

5. Python allows you to enter floating-point values in "calculator" notation, similar to scientific notation. For example, `1.23e5` represents 1.23×10^5; it will be displayed as `123000.0`. By default, Python also displays large `floats` in "calculator" notation. Find the smallest number and the largest number that will be displayed in calculator notation.

6.◆ Find the fraction with a denominator less than or equal to 20 that best approximates π.

    ⋛ Hints:

    1.
       ```
       from math import pi
       ```

    2. Recall that the built-in function `abs` returns the absolute value of a number.

    3.
       ```
       if approx < best_approx:
               best_approx = approx
               best_num, best_denom = num, denom
       ```

    ⋛

6.5 Irrational Numbers

Between integers and fractions, all the numbers should be covered, right? That's what Pythagoras and his disciples believed, about 2500 years ago. Around 550 BC, Pythagoras founded a small religious "brotherhood," made up of Greek immigrants, in the city of Croton on the shores of southern Italy. Pythagoreans made great contributions to geometry, and they were the first to study numeric relations in music: they figured out simple ratios of frequencies for pleasant combinations of musical sounds. They literally worshiped numbers. But, as it often happens, their own discoveries eventually challenged their beliefs.

Pythagoreans believed that any number is either a whole number or a *rational number* (that is, a *ratio*, a fraction with an integer numerator and denominator). Pythagoreans thought of numbers in terms of geometry, as ratios of lengths of segments. They knew how to find the hypotenuse of a right triangle with the given legs (using the Pythagorean theorem, of course). So, eventually, they had to ask: what is the ratio of the diagonal of a square to its side? From the Pythagorean theorem, if the side is *a*, and the diagonal is *d*, we should have $a^2 + a^2 = d^2$ or $\frac{d^2}{a^2} = \left(\frac{d}{a}\right)^2 = 2$. So what kind of number is $\frac{d}{a}$? Is it a rational number? (Legend has it that when one of the disciples proved that this ratio cannot be a rational number, things didn't end well: fellow Pythagoreans drowned him in a fountain for this "heresy.")

The first proof of the irrationality of $\sqrt{2}$ most likely used geometry. Something like this. Suppose the side of a square is *mx* and the diagonal is *nx*, where *m* and *n* are integers and *x* is a "unit segment," a "common measure" of *a* and *d*. Let's assume that *m* and *n* are the smallest possible such integers. However, we can construct a smaller square (Figure 6-3) with side $(n - m)x$ and diagonal $(2m - n)x$ that has the same property: *n-m* and *2m-n* are integers. But these integers are smaller than *m* and *n*, respectively. So the initial hypothesis was wrong. This method of proof is called *proof by contradiction*.

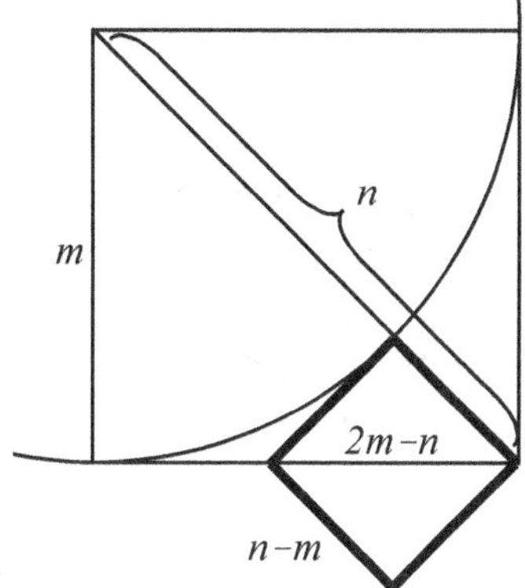

Figure 6-3. A geometric proof of the irrationality of $\sqrt{2}$.

Here is a more modern, algebraic version of the proof. Suppose there exists a rational number $\dfrac{n}{m}$ such that $\dfrac{n^2}{m^2} = 2$. Let's take the smallest such numbers m and n (by reducing the fraction). $n^2 = 2m^2$, so n must be an even number. (If n were odd, n^2 would be odd, too.) So $n = 2k$, for some k . Then $n^2 = 4k^2 \Rightarrow 4k^2 = 2m^2 \Rightarrow 2k^2 = m^2$. So m must be even, too. Since both m and n are even, we can reduce the fraction further and get a smaller integer numerator and denominator. This contradicts the assumption that m and n exist in the first place.

That's how the first *irrational number* was discovered. $a + b\sqrt{2}$ (with integers a and b), $\sqrt{3}$, and so on are all irrational, too. Eventually mathematicians proved that π is irrational. Much later, in the 19th century, the German mathematician Georg Cantor proved that there are "significantly more" irrational numbers than rational numbers. Both sets are infinite, of course. Cantor devised a way to compare infinite sets. In his theory, the set of rational numbers is *countable*, that is, it belongs to the same category of infinity as the set of positive integers. The set of irrational numbers is in a different category of infinity: it is not countable.

Section 6.5 ~ Exercises

1.   Positive integers a, b, and c are said to form a *Pythagorean triple* if $a^2 + b^2 = c^2$. It is fairly easy to show that if we take any integers $0 < m < n$, then $a = n^2 - m^2$, $b = 2mn$, and $c = n^2 + m^2$ form a Pythagorean triple. For example, if we take $m = 1$, $n = 2$, we get $a = 3$, $b = 4$, $c = 5$. How many different pairs of integers $0 < m < n \le 7$ are there? Write a Python program that will generate and display Pythagorean triples for all such pairs of m and n. ✓

2. ▪   If $a = \sqrt{2}$, then $a = \dfrac{2}{a}$. Let's start with $x = 1$ and repeatedly replace x with the arithmetic mean of x and $\dfrac{2}{x}$. If we keep repeating this operation, x will get closer and closer to $\sqrt{2}$. Write a program that uses this method to evaluate $\sqrt{2}$ accurately to at least 5 digits after the decimal point. This accuracy is achieved when $\left| x - \dfrac{2}{x} \right| < 0.00001$. Display the resulting estimate and the number of iterations that was required to obtain it. Compare your result with the value returned by `sqrt(2)`.

3. ▪   The golden ratio $\varphi = \dfrac{1 + \sqrt{5}}{2}$. Find the fraction with a denominator under 50 that best approximates the golden ratio.

6.6 Review

Terms and notation introduced in this chapter:

Positional number system *Two's complement representation*
Base of a number system   *of negative integers*
Binary number system   *Floating-point number*
Octal number system   *Rational number*
Hexadecimal number system *Irrational number*

Some of the Python features introduced in this chapter:

Binary constants and conversions:

```
int('0010110', 2)
'{0:08b}'.format(38)
bin(38)
0b100110
```

Octal constants and conversions:

```
oct(38)
'{0:04o}'.format(38)
int('46', 8)
0o46
```

Hexadecimal constants and conversions:

```
hex(466)
'{0:04x}'.format(466)
int('12AB', 16)
0x12AB, 0x12ab
```

if chapter is 7:

Boolean Algebra and
`if-else` Statements

7.1 Prologue

A *proposition* is a statement that is either true or false. For example: "2020 is a leap year," or "100 is a prime." Formal logic deals with the laws of reasoning that apply to propositions. Logical operations on propositions are not unlike arithmetic operations on numbers. In arithmetic, we can express common properties of operations using algebraic notation. For example, take the distributive law for multiplication:

$$a(b + c) = ab + ac$$

Similarly, we can use the notation of *Boolean algebra* to express general laws of logic. For example:

$$not\ (A\ and\ B) \Leftrightarrow (not\ A)\ or\ (not\ B)$$

$P \Leftrightarrow Q$ means "P is equivalent to Q" or "P is true if and only if Q is true." ("If and only if" is a common phrase in mathematics. It is analogous to "is equal" for numbers.) The above law, for example, says: *not* (*rich and famous*) is equivalent to (*not rich*) *or* (*not famous*).

Many mathematicians use a special notation for logical operations:

$\neg$ means "not"

$\wedge$ means "and"

$\vee$ means "or"

In this notation, the above formula would look like this:

$$\neg(P \wedge Q) \Leftrightarrow (\neg P) \vee (\neg Q)$$

In this book, we will continue using *and, or, not* for the sake of simplicity, and because `and`, `or`, and `not` are Python keywords.

Boolean algebra is named after the British mathematician George Boole (1815-1864), who described the laws of logic in his books *The Mathematical Analysis of Logic* and *The Laws of Thought*.

7.2 Operations in Boolean Algebra

If you have two propositions, P and Q, the proposition "*P and Q*" is called their *conjunction*. For example, the conjunction of two propositions

> *I like coffee*
> *you like tea*

is the proposition

> *I like coffee <u>and</u> you like tea*

> **The proposition *P and Q* is true if and only if both *P* is true <u>and</u> *Q* is true.**

Conjunction (the *and* operation) can be described by the following table (where T stands for "true" and F stands for "false"):

P	Q	$P\ and\ Q$
T	T	T
T	F	F
F	T	F
F	F	F

A table of this kind is called a *truth table*. It shows the values of a Boolean (logical) expression (in this case, $P\ and\ Q$) for all possible combinations of values of the variables in it.

❖   ❖   ❖

"*P or Q*" is called the *disjunction* of P and Q.

For example, the disjunction of the propositions

> *we can dance*
> *we can sing*

is the proposition

> *we can dance <u>or</u> we can sing* (either one or both)

> **The proposition *P or Q* is true if and only if <u>at least one</u> of the propositions *P* and *Q* is true, that is, *P* is true <u>or</u> *Q* is true or <u>both</u> are true.**

Disjunction (the *or* operation) has the following truth table:

P	*Q*	*P or Q*
T	T	T
T	F	T
F	T	T
F	F	F

If *P* is a proposition, "*not P*" is called its *negation*. For example, "100 is not a prime" is the negation of "100 is a prime."

> ***not P* is true if and only if *P* is false.**

Negation (the *not* operation) has the following truth table:

P	*not P*
T	F
F	T

> **A *Boolean expression* is a correctly formed formula that involves variables, *and*, *or*, and *not* operators, and parentheses. The value of a Boolean expression is true or false.**

In English, the words "and," "or," "not" can be used differently in different contexts. For example, the statement "Henry likes macaroni and cheese" most likely means that Henry likes macaroni <u>with</u> cheese, not that he likes macaroni and also likes cheese. Also, in English, "or" is often used as *exclusive or*: *P* or *Q*, but not both. For example: "Live free or die" means do one or the other, but not both. In Boolean algebra, however, the operations *and*, *or*, and *not* have precise meanings, as described in their truth tables above.

Table 7-1 lists some of the important laws of logic.

not (not P) $\Leftrightarrow$ P	Double negation law
P or (not P) $\Leftrightarrow$ T	Law of the excluded middle
P and (not P) $\Leftrightarrow$ F	Law of contradiction
P and Q $\Leftrightarrow$ Q and P P or Q $\Leftrightarrow$ Q or P	Commutative laws
P and (Q or R) $\Leftrightarrow$ (P and Q) or (P and R) P or (Q and R) $\Leftrightarrow$ (P or Q) and (P or R)	Distributive laws
not (P and Q) $\Leftrightarrow$ (not P) or (not Q) not (P or Q) $\Leftrightarrow$ (not P) and (not Q)	De Morgan's laws

Table 7-1. Some of the laws of formal logic

Notice the *duality* of the laws: if you take any law of logic and replace each *and* with *or* and each *or* with *and*, you get another valid law of logic. In Table 7-1, dual laws are written in the same block.

Of all the laws in the table, De Morgan's laws are the least obvious. These laws are very useful in computer programming. We gave an example of one in the prologue.

In regular algebra, there is only one distributive law: $a(b+c) = ab + ac$. In Boolean algebra, there are two distributive laws.

It is much easier to prove the laws of Boolean algebra than the laws of algebra of numbers, because we can test all possible combinations of the variables involved and compare the truth tables of the expressions on the left and on the right. The truth table for a law with two variables has four rows; the truth table for an expression with three variables has eight rows.

Section 7.2 ~ Exercises

1.   What is the negation of "I do"?

2.   What is the conjunction of "I did my homework" and "I went to see a movie"? ✓

3.   Rephrase in English the negation of "neither here nor there"? ✓

4.   Restate the proposition "I can't read or write" as a Boolean expression in terms of P and Q, where P is "I can read" and Q is "I can write." Write the negation of your expression and simplify it using De Morgan's laws.

5.   Make a truth table for the expression P *or* (*not Q*).

6.■   Make a truth table for the expression P *and* (Q *or* R). ⸮ Hint: your table needs all possible combinations of T/F values for P, Q, and R, so it will have 8 rows. ⸮ ✓

7.■   Prove the De Morgan's law

$$\text{not } (P \text{ and } Q) \Leftrightarrow (\text{not } P) \text{ or } (\text{not } Q)$$

by showing that the truth tables for the left and right sides of this law are identical.

8.◆   Write a logical expression with two variables, P and Q, that has the following truth table:

P	*Q*	*<Your expression>*
T	T	F
T	F	T
F	T	T
F	F	F

⸮ Hint: write the expression as a disjunction of some of the conjunctions P *and* Q, P *and* (*not Q*), (*not P*) *and* Q, (*not P*) *and* (*not Q*), then simplify. ⸮ ✓

7.3 Logic and Sets

Suppose we choose a set and work only with its subsets. We will call this set the *universal set* and use the letter U for it. For different types of math problems, we might choose different universal sets. For example, we might make the universal set be all real numbers $U = \mathbb{R}$, or all integers $U = \mathbb{Z}$. Once the universal set U has been defined, we can consider a subset A of U that consists of all the elements of U that meet certain criteria or have a certain property. In other words, a certain statement p about an element x of U will be true if and only if x belongs to A:

$$x \in A \iff p(x) = \text{true}$$

p can be viewed as a function from U to the set {true, false}. In mathematical logic such a function is called a *predicate* on U. The subset of U where p is true is called the *truth set* of p (Figure 7-1).

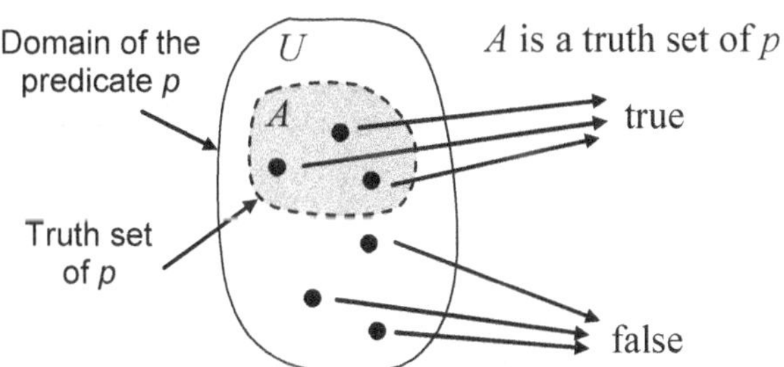

Figure 7-1. The truth set of a predicate

Example 1

Suppose our universal set U is the set of all real numbers $U = \mathbb{R}$ and the predicate $p(x)$ on U is "the square of x does not exceed 4." Then its truth set A_p is the interval $-2 \le x \le 2$.

Example 2

Suppose our universal set is the set of all positive integers. A positive integer is called a *Mersenne prime* if it is a prime (has no divisors except 1 and itself) and has the form $2^k - 1$ for some positive integer k. For example, 3, 7 and 31 are Mersenne primes. Let M be the set of all Mersenne primes. Then M is the truth set of the predicate "n is a prime and $n = 2^k - 1$ for some $k \geq 2$."

Example 3

Suppose we toss a coin three times and record the result as a sequence of three digits, using 0 for heads and 1 for tails. Let U be the set of all possible outcomes: $\{000, 001, 010, 011, 100, 101, 110, 111\}$. What is the truth set of the predicate "there were at least two tails in a row"? Now describe the negation of this predicate and its truth set.

Solution

The truth set of "there were at least two tails in a row" is $\{011, 110, 111\}$. The negation of "there were at least two tails in a row" is "there were not at least two tails in a row." Its truth set is $\{000, 001, 010, 100, 101\}$.

The connection between predicates and subsets of the universal set is so obvious that it appears to be a *tautology* (same thing in different words). Yet when we can establish an exact match between two mathematical theories, we usually can get some interesting results.

- The *intersection* of sets A and B (which are subsets of U) is the set of all the elements of U that belong to both A <u>and</u> B (see Figure 7-2-a).

- The *union* of sets A and B is the set of all the elements of U that belong to either A <u>or</u> B or both (see Figure 7-2-b).

- The *complement* of a set, $\overline{A} = U - A$, consists of all the elements of U that are <u>not</u> in A (see Figure 7-2-c).

The symbols for the intersection and union of sets, $\cap$ and $\cup$, are similar to the symbols for conjunction and disjunction in mathematical logic, $\wedge$ and $\vee$.

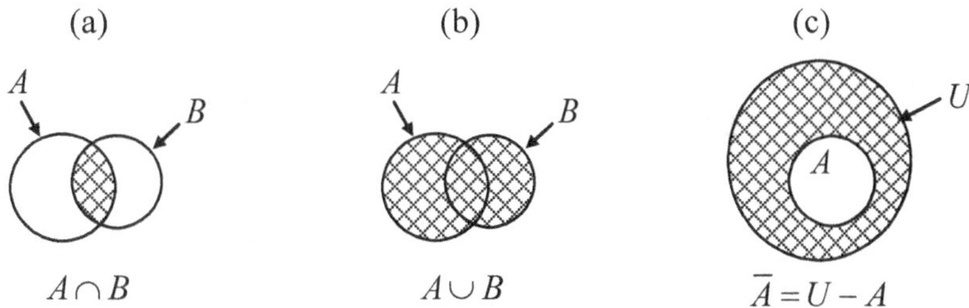

Figure 7-2. $A \cap B$, $A \cup B$, and $\overline{A} = U - A$

For example, the truth set for Mersenne primes in Example 2 is the intersection of the set of all primes and the set of all numbers equal to $2^k - 1$ for some k.

You are probably familiar with *Venn diagrams,* which illustrate graphically the parallelism of logic and sets. (Unsurprisingly, Venn diagrams are sometimes called *set diagrams* or *logic diagrams.*) The Venn diagram in Figure 7-3 shows how the logical operation *and* corresponds to the intersection of sets.

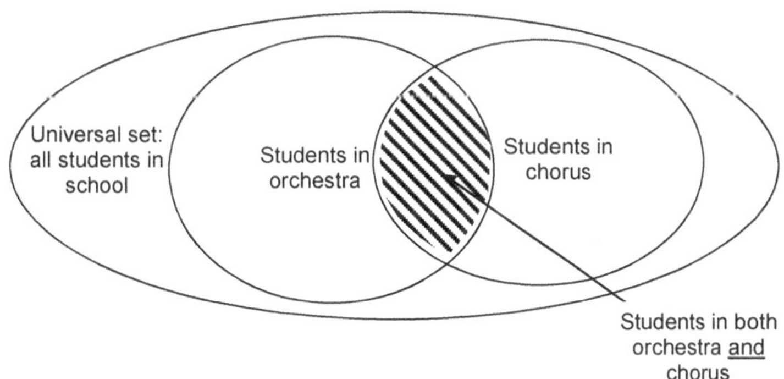

Figure 7-3. An example of a Venn diagram

Boolean algebra and the set theory are two mathematical theories that model each other. Each law of Boolean algebra has a corresponding law in the algebra of sets, and vice versa. Figure 7-4 shows the matching notations for operations in logic and sets.

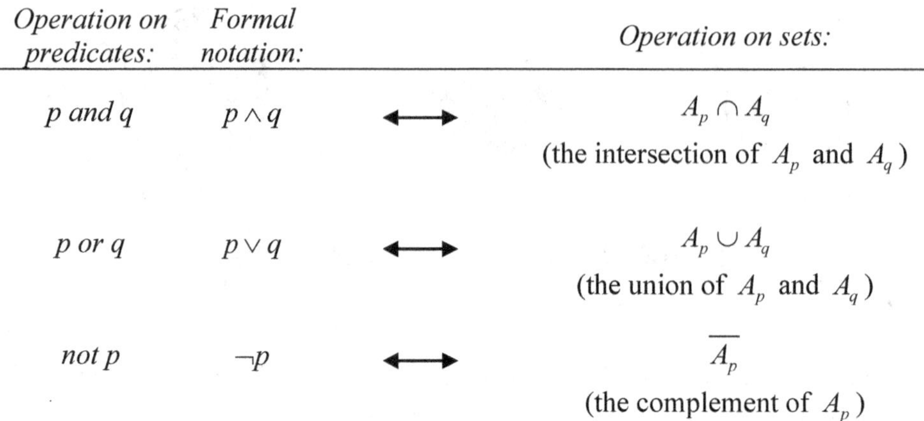

Operation on predicates:	Formal notation:		Operation on sets:
p and q	$p \wedge q$	$\longleftrightarrow$	$A_p \cap A_q$ (the intersection of A_p and A_q)
p or q	$p \vee q$	$\longleftrightarrow$	$A_p \cup A_q$ (the union of A_p and A_q)
not p	$\neg p$	$\longleftrightarrow$	$\overline{A_p}$ (the complement of A_p)

Figure 7-4. Notation for logic and sets

Example 4

One of the two De Morgan's laws in Boolean algebra is $\neg(P \wedge Q) \Leftrightarrow (\neg P) \vee (\neg Q)$. The corresponding law in set theory is $\overline{A \cap B} = \overline{A} \cup \overline{B}$.

Python has operators that work on sets:

- The `&` operator, when applied to two sets, gives their intersection.

- The `|` operator gives the union of two sets.

- The `-` operator gives the difference of two sets: `s1-s2` contains all the elements from `s1` that are not in `s2`.

- The `^` operator is called *symmetric difference*; `s1^s2` includes all the elements that are in `s1` or in `s2` but not in both.

Section 7.3 ~ Exercises

1. Some students do drama. Some do chorus. Some do band. Some do two of the above. Some do all three. Draw a Venn diagram that illustrates this situation.

2. Sketch a set of all points x on the number line such that $(x-1)^2 \geq 4$. ✓

3. Sketch a set of all points (x, y) on the x-y plane such that $x \geq 0$, $y \geq 0$, and $x^2 + y^2 \leq 1$.

4. Sketch a set of all points (x, y) on the x-y plane such that $(x-2)(y-1) \geq 0$. ✓

5. In a Sudoku puzzle, the objective is to fill a 9-by-9 grid with digits from 1 to 9 in such a way that each row, each column, and each of the nine 3-by-3 squares holds all the digits 1 through 9. The grid below shows a Sudoku grid with some of the 1s and 2s on it.

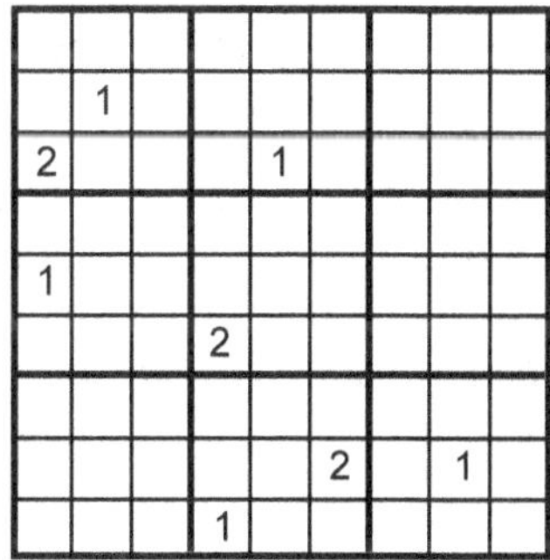

Mark all the remaining squares on the grid for which <u>both</u> 1 and 2 are possible candidates. ✓

6. In a study of health problems related to obesity, 34 percent of subjects were found to be obese, 14 percent had diabetes, and 12 percent both were obese and had diabetes. What percent of subjects either were obese or had diabetes (or both)? ✓

7.■   Suppose two predicates p and q (on the same set) are such that if $p(x)$ is true then $q(x)$ is true. In this case we say that p *implies* q and write $p \rightarrow q$. Describe the relationship between the truth sets of such predicates. ✓

8.   Translate all the laws of logic from Table 7-1 into the corresponding laws for sets.

9.   Prove one of the two De Morgan's laws for sets, using a Venn diagram.

10.■   Express the Python operator ^ for sets using only the &, |, and – operators. ✓

11.■   Fill in the blanks in this function, which takes two strings of uppercase letters and returns a sorted list of letters that are in the first string but not in the second.

```
def exclude_letters(word1, word2):
    """Return a sorted list of letters that are in word1
       but not in word2, without duplicates.
    """

    return ________________________________________
```

7.4 `if-else` Statements in Python

Programs often need to determine how to proceed based on a certain condition. CPUs have special *conditional jump* instructions that either continue with the next instruction or jump to a different place in the program, depending on the result of the previous operation (see the example in Section 1.2). High-level languages have "if-else" statements for conditional branching.

Example 1

An `if-else` statement in Python

Suppose the string s represents an integer, possibly a negative one, and we want to get hold of its sign. Then we can write:

```
if len(s) > 0 and s[0] == '-':
    sign = -1
else:
    sign = 1
```

This statement works as follows: if `s` is not empty and the first character in `s` is `'-'`, then `sign` gets the value `-1`; otherwise, `sign` gets the value `1`.

The general syntax for Python's `if-else` statement is

```
if <condition>:
    <statement 1A>
    <statement 1B>
    ...
else:
    <statement 2A>
    <statement 2B>
    ...
```

If the condition is true, the statements indented under `if` are executed; otherwise, the statements indented under `else` are executed. After that, the program proceeds with the statement that follows `if-else`. Don't forget the colons!

The `else` clause is optional: you can have `if` alone. In that case, if the condition is true, the program executes the statements indented under `if`; otherwise, the program skips them.

Example 2

if with no else

```
if x < 0:
    x = -x
...
```

If you want to "do nothing" under `if` and something under `else`, you can use Python's "do nothing" keyword, `pass`.

Example 3

if with pass

```
if y == 0:
    pass    # do nothing
else:
    print(1/y)
...
```

Often you need to chain several `if-else` statements together. For example:

```python
def letter_grade(score):
    if score >= 90:
        return 'A'
    else:
        if score >= 80:
            return 'B'
        else:
            if score >= 70:
                return 'C'
            else:
                if score >= 60:
                    return 'D'
                else:
                    return 'F'
```

Python lets you simplify the indentation and compress the "else-if" on one line, using the keyword `elif`. The above code can be shortened to:

```python
def letter_grade(score):
    if score >= 90:
        return 'A'
    elif score >= 80:
        return 'B'
    elif score >= 70:
        return 'C'
    elif score >= 60:
        return 'D'
    else:
        return 'F'
```

Python has two special built-in constants: `True` and `False`. A condition is a Boolean expression that evaluates to `True` or `False`. So a condition is basically a predicate on the values of the variables involved. We have already encountered conditions when we talked about `while` loops, in Chapter 4.

Conditions are written using *relational operators* and *logical operators*. Python has the following relational operators:

Operator	Description
`==`	equal
`is`	equal; same as `==`
`!=`	not equal
`is not`	not equal; same as `!=`
`<`	less than
`<=`	less than or equal to
`>`	greater than
`>=`	greater than or equal to
`in`	includes the value
`not in`	does not include the value

The relational operators also apply to strings, lists, and tuples. For example:

```
>>> 'Atlanta' < 'Boston'
True


>>> [1, 2, 3, 4] < [1, 3, 4]
True
```

Strings are compared alphabetically, but all the uppercase letters are smaller than any lowercase letter. Lists are compared via pairwise comparisons of elements with the same indices.

The conditions `x in s` and `x not in s` can be used with any "sequence" `s` (range, string, list, tuple, set).

Python's logical operators (`and`, `or`, and `not`) correspond to the logical operators of Boolean algebra. You can try out these operators in Python's shell. For example:

```
>>> 3 >= 4
False
>>> 3 < 5 and 3 > 1
True
>>> 3 < 5 or 3 < 0
True
```

The relational operators take precedence over the logical operators: they are applied first.

That is why we didn't need to use parentheses in the above examples.

Example 4

Suppose `name` is a string and `guests` is a list. Write a code segment that checks whether `name` is in the list `guests`, and if so, prints "Welcome," followed by `name`. If `name` is not in `guests`, the code should print "May I help you?"

Solution

```python
if name in guests:
    print('Welcome', name)
else:
    print('May I help you?')
```

❖   ❖   ❖

Conditions can also include calls to *Boolean functions*. (By Boolean function, we mean a function that returns either `True` or `False`.) Python has a small number of built-in Boolean functions (for example, `any` and `all`), and you can write your own.

Example 5

The function `is_prime` below returns `True` if *n* is a prime number; otherwise it returns `False`.

```python
from math import sqrt

def is_prime(n):
    """Return True if n is a prime number; otherwise return False."""
    if n < 2:
        return False
    d = 2
    while d <= round(sqrt(n)):    # if n is divisible by k, then n is
                                  #    also divisible by n/k and either
                                  #    k or n/k must be below sqrt(n)

        if n % d == 0:
            return False
        d += 1
    return True
```

Test it:

```
>>> is_prime(5)
True
>>> is_prime(6)
False
>>> n = 7
>>> n < 10 and is_prime(n)
True
```

You can return the value of a Boolean expression from a function.

Example 6

```
def greater_than(a, b):
    if a > b:
        return True
    else:
        return False
```

is the same as:

```
def greater_than(a, b):
    return a > b
```

Other examples of good and bad style:

Good	Verbose and redundant
`if name not in guests:`	`if name in guests == False:`

`found = name in guests`	`if name in guests:`
	`    found = True`
	`else`
	`    found = False`

Python also has a shortcut for expressions that evaluate to different values based on a condition. The syntax is:

```
a if <condition> else b
```

For example:

```
abs_x = x if x >= 0 else -x
```

❖   ❖   ❖

In logic there are commutative laws:

$$P \text{ and } Q \Leftrightarrow Q \text{ and } P$$
$$P \text{ or } Q \Leftrightarrow Q \text{ or } P$$

With computers, things are slightly more complicated: you have to pay attention to the order of the operands. When a program evaluates a Boolean expression, it actually performs certain operations. In the process it may encounter an error or raise an exception. For better efficiency and to allow more flexibility, many languages, including Python, obey *short-circuit evaluation* when evaluating Boolean expressions.

> **When evaluating p and q, Python first evaluates p. If p is False, then the result is False, and q is not evaluated.**

> **When evaluating p or q, Python first evaluates p. If p is True, the result is True, and q is not evaluated.**

Short-circuit evaluation breaks the symmetry of commutative operations. For example, if s happens to be an empty string in

```
if len(s) > 0 and s[0] == '-':
    sign = -1
else:
    sign = 1
```

then `len(s) == 0`, and the code will set `sign` to 1. But if we write

```
if s[0] == '-' and len(s) > 0:
    sign = -1
else:
    sign = 1
```

an empty string will raise the exception "IndexError: string index out of range", because the character `s[0]` does not exist.

Section 7.4 ~ Exercises

1. Write a Boolean expression in Python that says "*n* is a positive integer and an even number." ✓

2. Simplify using one of De Morgan's laws:

```
not (x >= -1 and x <= 1)
```

3. Which of the following expressions are equivalent to
`not (a or (not b))`? ✓

    (a)   `(not a) or (not b)`
    (b)   `(not a) or b`
    (c)   `(not a) and b`

4. Write a function that returns the number of days in a given month specified by name: `'January'`, `'February'`, and so on. Use an `if-elif-else` sequence. ✓

5. ▪ Write a function `is_leap_year` that returns `True` if a given year is a leap year. A year is a leap year if it is divisible by 4 but not by 100; however, if the year is divisible by 400, it is a leap year, too. For instance, 2000 and 2008 were leap years, but 1900 was not. ✓

6. Using the function `is_prime` from Example 5, write a program that prints all the prime numbers under 1000. To make your program more general, define a variable `n = 1000` and print all the prime numbers under *n*.

7. Write a Boolean expression in Python that evaluates to `True` if and only if the string `s` contains either a `'+'` or a `'-'`, but not both. ✓

8. ▪   Fill in the blank:

```
def is_hex_digit(d):
    """ Return True if d is a single character and d
        is a hex digit: '0' - '9', 'a' - 'f' or 'A' - 'F';
        otherwise return False.
    """

    return ______________________________________________
```

✓

9.   For which values of x will the following statement raise an exception?

```
if x >= 0 and 1.0 / sqrt(x) < 0.01:
    pass
```

10. ◆   Write a program that plays Rock-Paper-Scissors with the user and displays the cumulative score. For example:

```
Rock... Paper... Scissors... Shoot!
Make your move (r, p, s) or q (to quit): s

I said Rock
Ha! You are zapped -- 1:0

Rock... Paper... Scissors... Shoot!
Make your move (r, p, s) or q (to quit): p

I said Paper
Paper-aper! Tie -- 1:0

Rock... Paper... Scissors... Shoot!
Make your move (r, p, s) or q (to quit): q

Sorry, you lost! 1:0
Thanks for playing.
```

‭ Hint: `choice(s)` returns a randomly chosen character from the string s (or a randomly chosen element from the list or tuple s). The function `choice` is in the module `random`, so you need

```
from random import choice
```

11. Write a function `is_earlier(date1, date2)` that takes two valid dates and returns `True` if `date1` is earlier than `date2` and false otherwise. Assume that a date is represented by a tuple `(month, day, year)`. ✓

12. Suppose you already have a function `is_prime`, as shown in Example 5. Write a function `is_Mersenne_prime` (see Example 2 in Section 7.3). Do not duplicate the code from `is_prime` — just call it.

13. Python has several "Easter eggs," or funny surprises, hidden in its software. For example, enter

```
>>> import this
```

Then try to predict Python's responses in this dialog:[*]

```
>>> love = this
>>> this is love
___________
>>> love is True
___________
>>> love is False
___________
>>> love is not True or False
___________
>>> love is not True or False; love is love
___________
```

7.5 Review

Terms and notation introduced in this chapter:

Boolean algebra   *De Morgan's laws*
Proposition   *Predicate*
Conjunction   *Truth set*
Disjunction   *Intersection of sets*
Negation   *Union of sets*
"And" operation   *Complement set*
"Or" operation   *Venn diagram*
"Not" operation   *Relational operators*
Truth table   *Logical operators*

---

[*] Posted on https://github.com/OrkoHunter/python-easter-eggs and many other web pages.

Some of the Python features introduced in this chapter:

Intersection, union, difference, and symmetric difference operators for sets: `&`, `|`, `-`, and `^`

`True, False` **keywords**

```
if < condition >:
    . . .
else:
    . . .
```

```
if < condition >:
    pass
else:
    . . .
```

```
if < condition >:
    . . .
elif < other condition >:
    . . .
. . .
else:
    . . .
```

Relational operators `==, is, !=, is not, <, <=, >, >=, in, not in`

Logical operators `and, or, not`

```
return < Boolean expression >
```

Chapter |= 0x08

Digital Circuits and Bitwise Operators

8.1 Prologue

The circuit in Figure 8-1 shows a light controlled by an on-off switch: when the switch is closed, it turns on the light. If we put two switches in a row (Figure 8-2-a), both Switch 1 <u>and</u> Switch 2 must be closed to turn on the light. If we put the switches parallel to each other (Figure 8-2-b), Switch 1 <u>or</u> Switch 2 (or both) must be closed to turn on the light. Sounds familiar?

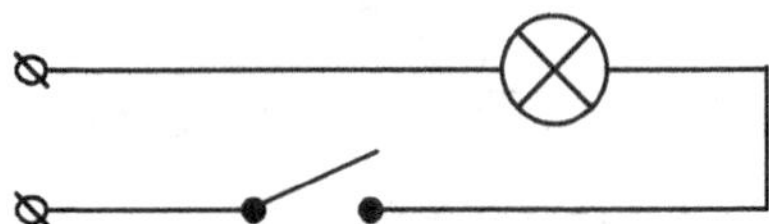

Figure 8-1. A circuit with a light controlled by one switch

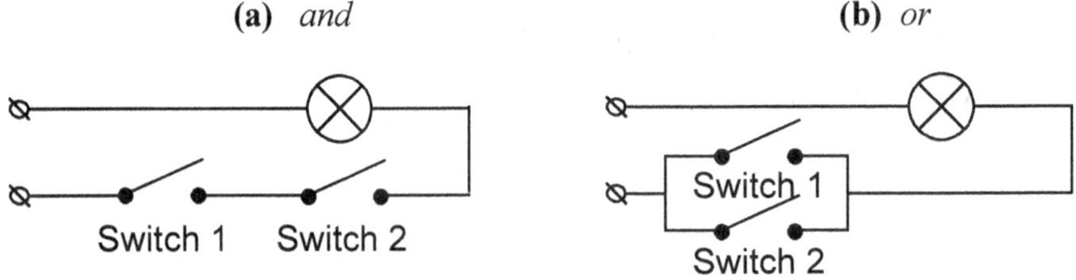

Figure 8-2. Arrangements of switches that model logical operations

The switches in Figure 8-1 and Figure 8-2 are manual switches. A *relay switch* has an electrical magnet controlled by another current: when a current runs through the magnet, the magnet pulls in the lever that closes the switch (Figure 8-3). This way, one current can control another current.

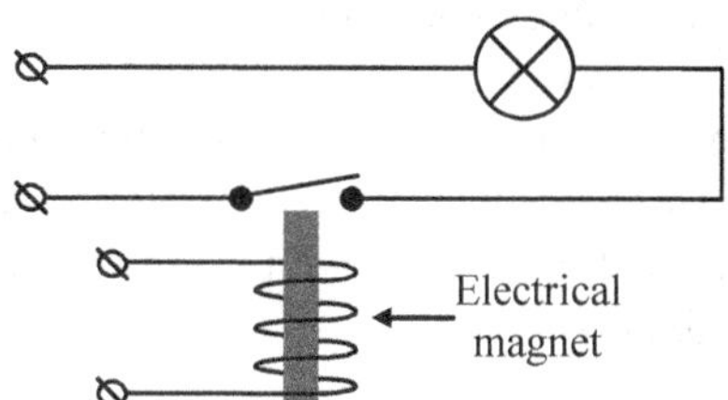

Figure 8-3. A relay switch

If you put two relay switches in a row, you will get an AND circuit (Figure 8-4-a). Two relay switches in parallel will make an OR circuit (Figure 8-4-b). You can also make a NOT circuit by using a special relay switch, which is normally closed but opens when a current is applied to the magnet (Figure 8-5).

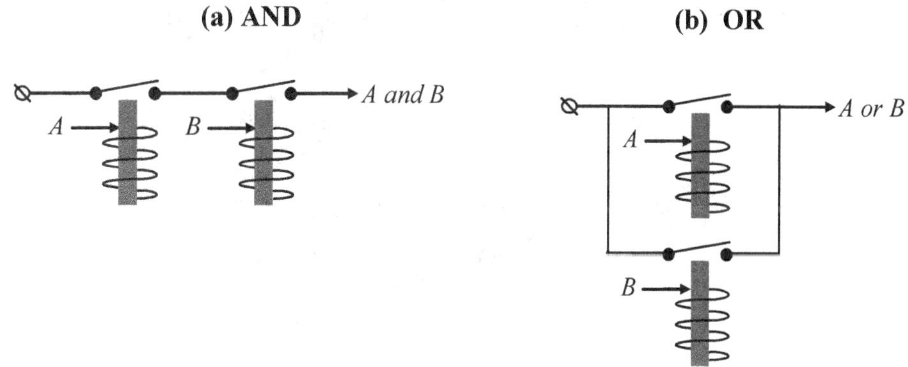

Figure 8-4.   A circuit made of two relay switches:
(a) AND (b) OR

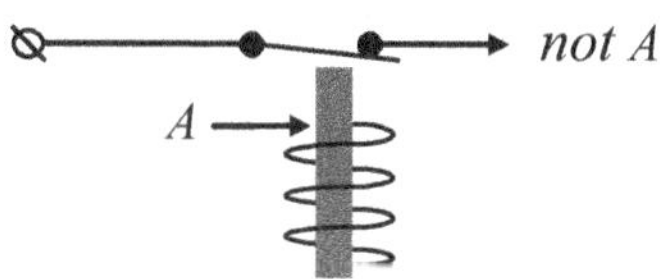

Figure 8-5. NOT circuit

AND, OR, and NOT circuits can be combined into more complex circuits. The output of one circuit serves as input to another.

In the late 1940s, John von Neumann, a great mathematician and one of the pioneers of computer technology, showed that any computation can be accomplished by combining AND, OR, and NOT circuits. So a computer can be built out of simple relay switches. In fact, some early computers were built that way (Figure 8-6).

Figure 8-6. Harvard Mark II relay computer (1947)
Courtesy of IEEE Annals of History of Computing

In modern *digital electronics*, microscopic transistors play the role of relay switches. Millions of transistors are etched into a small silicon chip. AND, OR, NOT, and other simple circuits are called *gates*. This type of circuitry is called *digital electronics* (as opposed to *analog electronics*), because only the presence or absence of a signal (electrical current) matters: "on" or "off," 0 or 1. The sizes (amplitudes) of the currents are the same — they are not used to carry information.

8.2 Gates

In diagrams of digital circuits, each type of gate is designated by a different shape (Figure 8-7). The table under each gate shows its output for all combinations of inputs. These tables are, of course, the same as the truth tables for the AND, OR, and NOT operations in Boolean algebra. Here 1 stands for True and 0 stands for False.

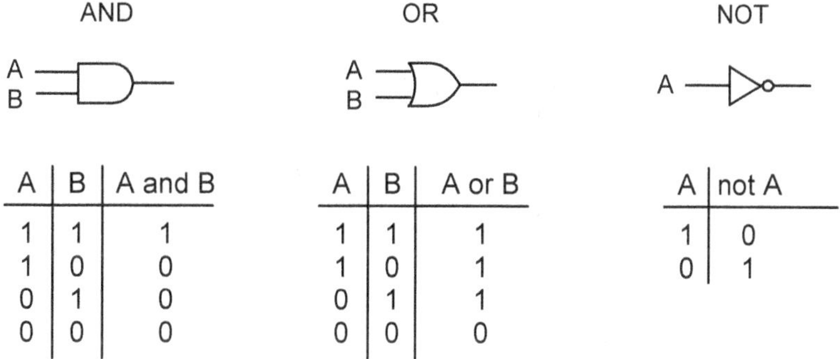

A	B	A and B
1	1	1
1	0	0
0	1	0
0	0	0

A	B	A or B
1	1	1
1	0	1
0	1	1
0	0	0

A	not A
1	0
0	1

Figure 8-7. AND, OR, and NOT gates and their outputs

Gates are combined to produce more complex circuits.

Example 1

The circuit

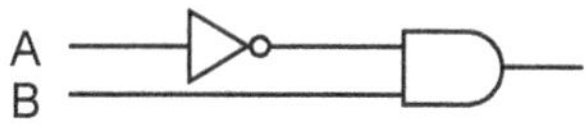

computes the logical expression `(not A) and B`. The output is defined by the following truth table:

A	B	(not A) and B
1	1	0
1	0	0
0	1	1
0	0	0

Example 2

The circuit

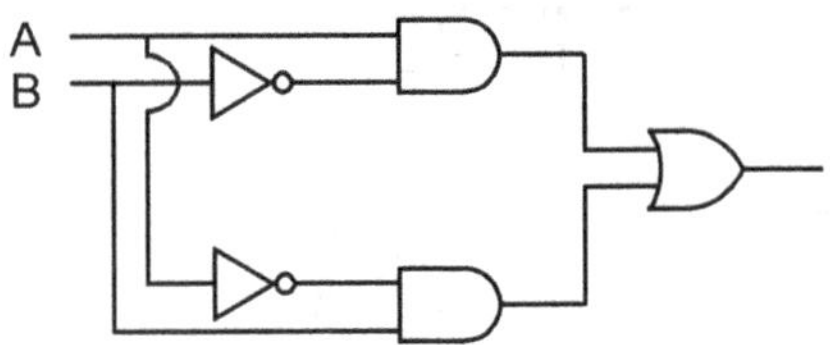

computes the logical expression

```
(A and (not B)) or ((not A) and B)
```

This is one of the ways to represent the *exclusive OR (XOR)* operation: the output is 1 if A is 1 and B is 0 or B is 1 and A is 0:

A	B	A xor B
1	1	0
1	0	1
0	1	1
0	0	0

❖   ❖   ❖

Real digital circuitry uses more types of gates: simple combinations of AND, OR, and NOT gates are combined into "compound" gates as follows:

Gate name	Gate symbol	Is short for:
XOR		
NAND		
NOR		
XNOR		

Section 8.2 ~ Exercises

1. Draw an electrical circuit with three on-off switches, such that closing any one of them turns the light on. ✓

2.■ Suppose you have several "two-way" switches at your disposal . A two-way switch can toggle a wire between two other wires:

Design an electrical circuit that allows you to turn a light on and off from two different locations (for example, you can turn on the light from a switch at the door, then turn off the light from another switch by your bed, then turn the light on again from either switch).

3. ▪   You have a battery-operated toy railroad and you want to be able to run a train forward and backward. The train's motor changes its direction of rotation if you switch the wires that go to the + and – contacts on the battery. You have a double two-way switch that can shift two wires from one pair of contacts to another:

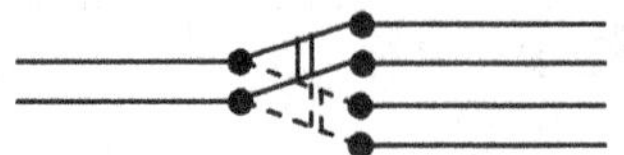

Design a circuit that lets you reverse the direction of the train by flipping the switch. ✓

4.   Using only AND, OR, and NOT gates, draw a circuit with two inputs, *A* and *B*, that computes *A or (not (A or B))*.

5.   Which of the gates listed in this section has the following truth table? ✓

A	*B*	output
1	1	1
1	0	0
0	1	0
0	0	0

6.   Using only AND, OR, and NOT gates, design a circuit with the following truth table:

A	*B*	output
1	1	0
1	0	0
0	1	0
0	0	1

Propose two designs: one using one AND and two NOT gates, the other using just two gates.

7.   Using only AND, OR, and NOT gates, draw a circuit with three inputs whose output is 1 if and only if all three inputs are 1. ✓

8. ▪   The circuit in Example 2 implements the XOR operation. Design an equivalent XOR circuit (with the same truth table) using four gates: one OR gate, two AND gates, and one NOT gate. ✓

9. Using only AND, OR, and NOT gates, design an XNOR circuit, with the following truth table:

A	B	output
1	1	1
1	0	0
0	1	0
0	0	1

Use no more than four gates total. ✓

10. Using AND, OR, and NOT gates, design a circuit with three inputs whose output is 1 if and only if at least two inputs are 1.

11. A "full adder" circuit has three inputs (the *A* bit, the *B* bit, and the carry bit) and two outputs (the result bit and the new carry bit). Design a full adder circuit using AND, OR, and NOT gates.

Several adders can be strung together to perform addition on binary numbers:

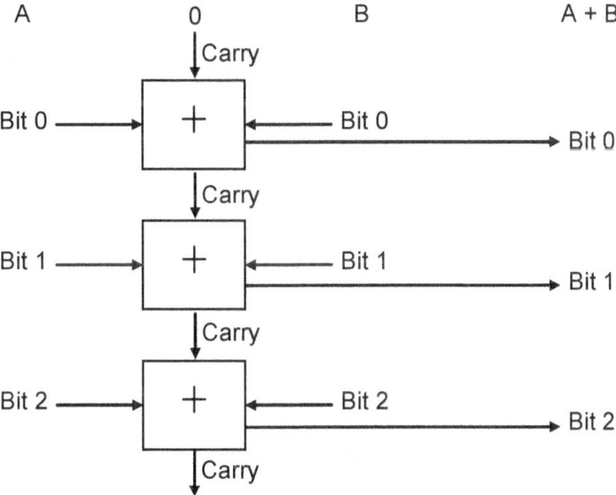

12. Construct a "comparator" circuit with four inputs that represent two 2-bit unsigned binary numbers $A_1 A_0$ and $B_1 B_0$. The output should be 1 when the first number is greater than the second; otherwise it should be 0.

8.3   Bitwise Logical Operators

As we know, memory locations and CPU registers in a computer can hold binary numbers, combinations of 0s and 1s stored in individual bits. A typical CPU provides instructions that perform the logical operations AND, OR, and NOT on respective bits in the operands. (An operand can be one byte, two bytes, four bytes or eight bytes.) A typical CPU also has the XOR instruction, which calculates the exclusive OR of the two operands. Figure 8-8 gives examples of the bitwise AND, OR, XOR, and NOT operations.

```
        01010110              01010110              01010110
AND  11110000          OR  11110000          XOR  11110000          NOT  01010110
     ========              ========              ========               ========
        01010000              11110110              10100110              10101001
```

Figure 8-8. Examples of bitwise AND, OR, XOR, and NOT operations

High-level languages usually provide bitwise logical operators that translate more or less directly into the respective CPU instructions. Python has the following bitwise logical operators:

Operator	Meaning
&	AND
|	OR
^	XOR
~	NOT

These operators work with `int` and Boolean operands.

(Recall that in Python &, | and ^ also give the intersection, union, and symmetric difference for two sets.)

In comments and text-mode rendition of math, x^n is often used to represent x to the power of n; in computer code, however, ^ represents XOR.

Recall that in Python, you can use the `int` function to convert a string of 0s and 1s into an `int`:

```
>>> int('11001110', 2)
206
```

The reverse operation is accomplished by the built-in function `bin`: `bin(n)` gives the string of binary digits of n, with the prefix `'0b'`. For example:

```
>>> bin(206)
'0b11001110'
```

Use `bin(n)[2:]` to get rid of the `'0b'` prefix. For example:

```
>>> bin(206)[2:]
'11001110'
```

Or use the `format` method of string to format the string of binary digits into a string of the given width, padded with zeros on the left. For example:

```
>>> '{0:012b}'.format(206)
'000011001110'
```

Or use the `rjust` method of string. For example:

```
>>> bin(206)[2:].rjust(12, '0')
'000011001110'
```

Python's bitwise negation operator ~ is not very useful, because it gives a negative result for a positive integer. Use ^ with a mask of "all ones" of a desired length instead. For example:

```
>>> bin(0b00001101 ^ 0b11111111)
'0b11110010'
```

> **Python does not give you an easy way to print out all the bits in a binary representation of a negative number.**

The `&` operator can be used to test whether a certain bit is set in a number, and the `|` operator can be used to set a bit. The hex constants

```
0x0001, 0x0002, 0x0004, 0x0008, 0x0010, 0x0020, 0x0040, 0x0080
```

correspond to bits 0, 1, 2, 3, 4, 5, 6, and 7, respectively.

Example 1

Test whether bit 2 is set in the binary representation of `error_code`:

```
bit = 0x0004
if (bit & error_code) != 0:
    ...  # do something
```

❖　❖　❖

Example 2

When we say "set a bit," we mean "set the bit to 1." Set bit 4 in `error_code`:

```
bit = 0x0010
error_code |= bit # the same as error_code = error_code | bit
```

❖　❖　❖

The `&` operator can be also used to "cut out" certain bits from a number, that is, to leave only the values of these bits and set the remaining bits to 0.

Example 3

Cut out the 4 least-significant bits from `c`. For example, if `c` has the value `0x482d`, then `byte0` should get the value `0x000d`.

Solution

```
mask = 0x000f
byte0 = c & mask
```

Example 4

Test whether bits 0-3 in `code` hold `1011`.

Solution

```
if (code & 0x0f) == 0x0b:
    ... # do something
```

Example 5

Test whether bits 0-7 in `pixels` are all set.

Solution

```
mask = 0xff
if (mask & pixels) == mask:
    ...
```

❖ ❖ ❖

In addition to bitwise logical instructions, CPUs have instructions that shift all the bits in an operand by a given number of positions. Python, too, has shift operators: >> and <<. For example:

```
>>> x = int('11001110', 2)
>>> bin(x)
'0b11001110'
>>> bin(x >> 1)
'0b1100111'
>>> bin(x << 1)
'0b110011100'
```

> **Shifting all the bits to the left by 1 is equivalent to multiplying the number by 2. Shifting all the bits to the right by 1 is equivalent to dividing the number by 2 (truncating the result to an integer).**

In other words, `a << 1` is the same as `a *= 2`, and `a >> 1` is the same as `a //= 2`.

Example 6

Write a function that returns the number of bits set to 1 in a given positive integer.

Solution

```
def count_bits(n):
    count = 0
    while n != 0:
        if n & 0b01 != 0:
            count += 1
        n >>= 1
    return count
```

Or, a little shorter:

```
def count_bits(n):
    count = 0
    while n != 0:
        count += n & 1
        n >>= 1
    return count
```

❖   ❖   ❖

Section 8.3 ~ Exercises

1. If `a` has the binary digits `01100111` and `b` has the binary digits `11010110`, write the result of `a & b`, `a | b`, `a ^ b`, `a << 1`, and `b >> 2`. ✓

2. Evaluate

    ```
    0xf8 & 0x8f
    0xf8 | 0x8f
    0xf8 ^ 0x8f
    ```

3. When a printer runs out of paper, bit 5 ("PE" — paper end) and bit 3 ("ERROR") in the 8-bit printer status register are both set to 1. Write a code fragment that tests whether the value `status_reg` indicates that the printer is out of paper. ✓

4. Fill in the blank in the following function without using the `%`, `//`, or `/` operators or strings. ✓

```
def divisible_by8(n):
    """Return True if n is divisible by 8;
       otherwise return false.
    """

    return ______________________________________
```

5. Write a statement that replaces bits 0-23 of `pattern` with their negatives, that is, replaces all 0s with 1s and 1s with 0s in those bits. The other bits in `pattern` should remain unchanged. ✓

6. If `a` is equal to `0x00c5`, what are the hex digits of `a | (a >> 1)`?

7. Write a Python statement that tests whether any of the bits 1, 3, or 5 is set in `byte`. ✓

8. Fill in the blanks in the function below, which returns a string of the eight hex digits (padded with zeros on the left, if necessary) of any `int` value n such that $0 \le n < 2^{32}$.

```
def my_hex(n):
    """Return the string of eight hex digits of n."""
    hex_digits = '0123456789abcdef'
    s = ''
    for k in range(______):

        s = hex_digits[n & ____________] + s

        n >>= __________

    return s
```

9.■ Write and test a function that takes an `int` value $0 \le n < 2^{32}$ and returns a new `int` with bits 0-29 in reverse order. The other bits should remain unchanged. Use only one loop, no `if`s, and only bitwise logical operators and shifts. ✓

10.   A tic-tac-toe position is represented in 27 bits (bits 0-26), three bits for each square. 000 means an empty square, 001 means an X, and 010 means an O. The triplets of bits go in order from left to right; first the top row, then the middle row, then the bottom row. The top left square is in bits 24-26; the bottom right corner is in bits 0-2. For example,

<pre>
  | O |
  | X |
X | X | O
</pre>

is represented as

000010000000001000001001010

Write a code fragment that checks whether there is an X in the central square.

11.♦   Write a function that tests whether a tic-tac-toe position, as described in Question 10, has a win for X (three X's in a row, including diagonals).

12.■   Find out whether the & and | operators can be applied to Boolean expressions and, if so, whether they use short-circuit evaluation. ✓

13.◆ Bits 0-23 in a positive integer represent a line of pixels (picture elements) in a black and white image (the remaining bits are 0s). 1 stands for black and 0 stands for white. We want to fill all single-pixel holes in the line, so that each pixel with black left and right neighbors becomes black. For example,

```
011110111000011011010000
```

should become

```
011111111000011111110000
```

Fill in the blank in the following function that does this and returns an `int` corresponding to the new line of pixels (with bits 0-23 updated as necessary and other bits set to 0).

```
def fill_holes(pix):

    return _________________________________________
```

14.◆ Count the number of *bitstrings* (strings of 0s and 1s) of length n that do not have two 1s in a row. First do it manually for $n = 1, 2, 3, 4, 5$; then fill in the blanks in the function below that counts such bitstrings for any $1 \le n \le 24$, and display this function's return values for n from 1 to 24.

```
def count_bitstrings(n):
    """Return the number of strings of 0s and 1s of length n
       that do not contain two 1s in a row.
    """
    return sum(1 for _________________________________________

                          if _________________________________)
```

Compare the output with your manual counts for small *n*.

8.4   Review

Terms introduced in this chapter:

> *Digital circuit*
> *AND, OR, and NOT gates*
> *XOR (exclusive OR)*
> *Bitwise logical operators*
> *Shift operators*

Some of the Python features introduced in this chapter:

> `&`, `|`, and `^` operators
>
> `>>` and `<<` operators

Chapter 9

Turtle Graphics

9.1 Prologue

Alice thought to herself, "I don't see how he can *ever* finish, if he doesn't begin." But she waited patiently.

"Once," said the Mock Turtle at last, with a deep sigh, "I was a real Turtle."

Alice's Adventures in Wonderland by Lewis Carroll

The idea of using computers and robots for teaching young children arose over half a century ago. In the late 1960s, three researchers, Wally Feurzeig and Cynthia Solomon from the research firm Bolt, Beranek and Newman (BBN), and Seymour Papert, a fellow at the Artificial Intelligence (AI) lab at the Massachusetts Institute of Technology (MIT), designed a first programming language for children. They called their language Logo, from the Greek word "logos," which means "word" or "thought." In those days, computers were big and expensive and used only for "serious" applications (military, data processing, research); to many people the idea of kids using valuable computer time sounded crazy. Yet Logo thrived, and within a few years it became popular among teachers and was introduced in many schools.

At first, Logo was meant to introduce young kids to AI ideas and methods. But one of Logo's features was a *virtual* (not physically existing) robot that could follow simple commands and draw pictures on the computer screen. Papert's group called the robot a "turtle" in honor of earlier "turtle" robots created by Grey Walter in the late 1940s (Figure 9-1). (The name "turtle" was reportedly inspired by the Mock Turtle character in Lewis Carroll's *Alice in Wonderland*.)

A real turtle robot that executed Logo instructions was built at MIT in 1969. In 1972, BBN engineer Paul Wexelblat designed and built the first wireless floor turtle (Figure 9-2).

Logo's "turtle graphics" capability quickly overshadowed Logo's other features, and it became known primarily as the turtle graphics language. Logo is alive and well today: many Logo versions and apps exist as free downloads, and turtle graphics ideas are implemented in other graphics packages and programming languages such as Scratch and, of course, Python's turtle graphics *module* (library of functions).

Figure 9-1.   A reproduction of one of Grey Walter's "turtle" robots

Courtesy http://roamerrobot.tumblr.com/post/23079345849/the-history-of-turtle-robots

Figure 9-2.   Paul Keelboat's wireless turtle, 1972

Courtesy http://cyberneticzoo.com/cyberneticanimals/
1969-the-logo-turtle-seymour-papert-marvin-minsky-et-al-american/

9.2 The `turtle` Module Basics

Python's `turtle` module comes with the standard Python installation from python.org. If you want to use `turtle`, you need to import it into your program:

```
from turtle import *  # import everything from the turtle module
```

If you wish, you can experiment with `turtle` commands (functions) directly from the Python shell. Try this:

```
>>> from turtle import *
>>> shape("turtle")
>>> forward(100)
```

A window will pop up with a line drawn by the turtle:

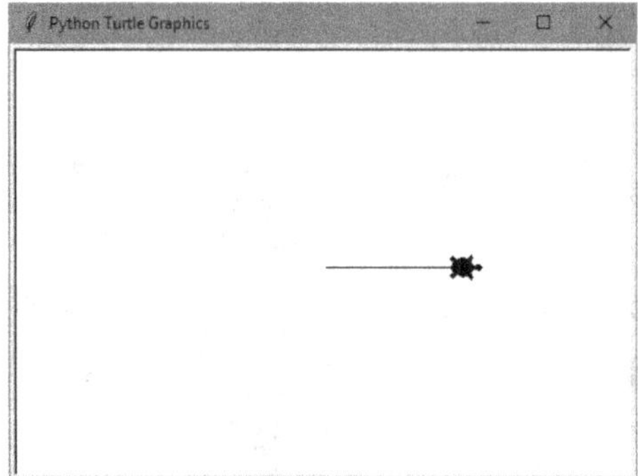

Table 9-1 shows the basic turtle commands (functions) needed to get you started. https://docs.python.org/3.7/library/turtle.html describes all the `turtle` and `screen` functions.

`turtle` functions are convenient and easy to use. Several names may be supported for the same function: a fully spelled-out name and an abbreviated name, such as `fd` for `forward`.

We will use full names of `turtle` functions in our programs for better readability and recommend you do the same — no need to save a few keystrokes.

Function	Action
`shape(name)`	Choose turtle's shape: `'arrow'`, `'turtle'`, `'circle'`, `'square'`, `'triangle'`, or `'classic'` (default). You can define your own shape.
`speed(v)`	Set turtle's moving and drawing speed: `'fastest'` or 0, `'fast'` or 10, `'normal'` or 6, `'slow'` or 3 (default), `'slowest'` or 1.
`color(colorname)`	Set pen and fill colors. `colorname` can be a string, for example, `color('red')`. (Other formats are supported — see Section 9.3.)

`penup()`
`pu()`
`up()`
	Lift the "pen" from the "paper" (ready to move without drawing).

`pendown()`
`pd()`
`down()`
	Place the "pen" on the "paper" (ready to draw).

`forward(d)` `fd(d)`	Move forward by `d` units (while drawing or not).

`backward(d)`
`bk(d)`
`back(d)`
	Move backward by `d` units.

`right(deg)` `rt(deg)`	Turn ↻ (clockwise) by `deg` degrees.
`left(deg)` `lt(deg)`	Turn ↺ (counterclockwise) by `deg` degrees.
`showturtle()` or `st()` `hideturtle()` or `ht()`	Make the turtle visible. Make the turtle invisible.

Table 9-1. Basic `turtle` functions

The distances in `turtle` functions are in *pixels* ("picture elements") by
default.

If the screen resolution in your device is listed, say, as 1920 by 1200, it means that the full screen is 1920 pixels horizontally and 1200 pixels vertically. The dimensions of the `turtle` graphics window are returned by the `window_width()` and `window_height()` functions. For example:

```
>>> from turtle import *
>>> window_width(), window_height()
(960, 900)
```

`setup(width, height)` defines your own custom size.

> **When a turtle is first created, it is placed at the center of the window, facing east (to the right), with its pen down, ready for drawing.**

> **If you have trouble figuring out turtle graphics code, imagine that *you* are the turtle and try following the commands. Just don't draw on the rug!**

The statement `from turtle import *` not only imports all `turtle` functions into your program, but also creates an anonymous `turtle` object whose functions can be called without any name-dot prefix. For example:

```
from turtle import *
shape('turtle')
speed('fastest')
forward(100)
```

You can create any number of other turtles and give them names. To call a named turtle's functions, you need to use the name-dot prefix. For example:

```
from turtle import *

setup(width=200, height=200)
alice = Turtle(shape='turtle')
alice.color('blue')
alice.forward(80)
bob = Turtle(shape='turtle')
bob.color('red')
bob.penup()
bob.right(90)
bob.forward(40)
bob.left(90)
bob.pendown()
bob.forward(80)
```

This displays

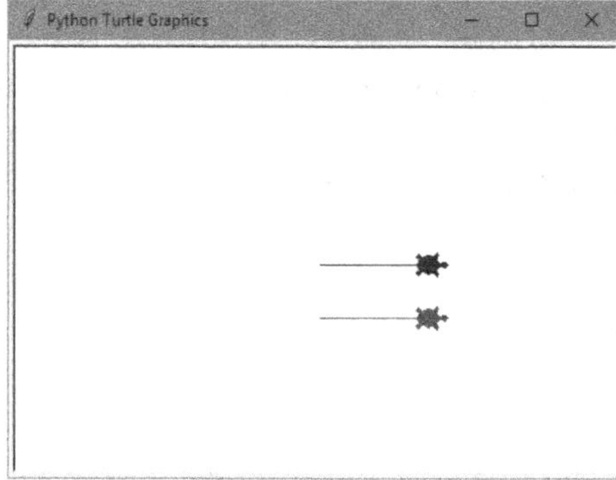

Example 1

Draw an equilateral triangle with a side length of 80 and a horizontal base centered in the middle of the graphics window:

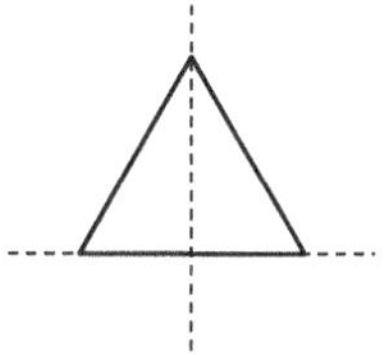

Solution

If the triangle is traced counterclockwise, the turtle needs to turn left by 120° after drawing each side:

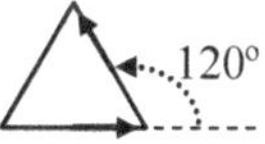


```
from turtle import *

penup()
backward(40)
pendown()
for k in range(3):
    forward(80)
    left(120)
```

Example 2

Draw a blue regular hexagon (a hexagon whose sides are all the same length and whose angles are all the same) that is centered at the center of the graphics window and has sides 100 pixels long. Return the turtle to the center of the window when done:

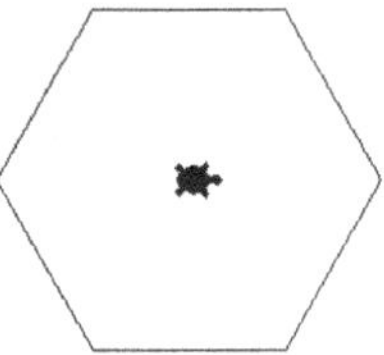

Solution

```
from turtle import *

shape('turtle')
color('blue')
penup()
backward(100)
right(60)
pendown()
for k in range(6):
    forward(100)
    left(60)
penup()
left(60)
forward(100)
```

Example 3

Write and test a function that draws a rectangle with given dimensions using a specified turtle, starting from that turtle's current position and direction.

Solution

```python
def draw_rectangle(t, w, h):
    """Draw a rectangle of width w and height h using the turtle t,
       going counterclockwise from its current position and direction
       and in its current color. Leave the turtle with its pen up.
    """
    t.pendown()
    for k in range(2):
        t.forward(w)
        t.left(90)
        t.forward(h)
        t.left(90)
    t.penup()
```

Then

```python
escher = Turtle()
escher.color('purple')
escher.left(45)
draw_rectangle(escher, 89, 55)
```

draws

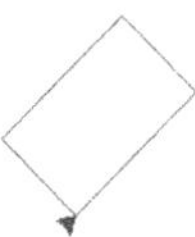

`turtle` can also draw filled shapes. Table 9-2 shows the relevant functions.

Function	Action
`begin_fill()`	Register the current position as the starting position for a filled shape.
`end_fill()`	End registering and fill the registered shape.
`color(c1, c2)`	Set the pen color to `c1` and the fill color to `c2`.

Table 9-2. Functions for filling shapes

The `begin_fill()` call tells the turtle to register the current position as the starting position for a filled shape. The `end_fill()` call fills the area within all the lines drawn since the `begin_fill` call. `t.color(c)` sets both `t`'s pen color and fill color to c, but you can specify different pen and fill colors by calling `color` with two parameters. For example,

```
>>> color('red', 'yellow')
```

sets the pen color to red and the fill color to yellow. The calls `pencolor(c)` and `fillcolor(c)` set the pen color and the fill color, respectively.

Example 4

Write and test a function `draw_chessboard(t, size, colors)` that uses the turtle `t` to draw a chessboard:

`size` is the size of each square; `colors` is a tuple of two colors, for the dark and light squares. Use the `draw_rectangle` function from the previous example.

Solution

```python
def draw_chessboard(t, size, colors):
    """Draw a chessboard using the turtle t with squares
       of a given size in given colors.
    """
    for row in range(8):
        for col in range(8):
            t.color(colors[(row + col)%2])
            t.begin_fill()
            draw_rectangle(t, size, size)
            t.end_fill()
            t.forward(size+1)
        t.back(8*size+8)
        t.right(90)
        t.forward(size+1)
        t.left(90)
```

```
# Draw the border:
t.back(1)
t.left(90)
t.forward(size)
t.right(90)
t.color('black')
draw_rectangle(t, 8*size+8, 8*size+8)

deepblue = Turtle()   # IBM's Deep Blue chess supercomputer beat Garry
                      # Kasparov, then the chess world champion, in 1997
deepblue.speed('fastest')
deepblue.hideturtle()   # to speed up the drawing
deepblue.penup()
deepblue.back(200)
draw_chessboard(deepblue, 40, ('yellow', 'blue'))
```

Or use the default anonymous turtle:

```
speed('fastest')
hideturtle()
penup()
back(200)
draw_chessboard(getturtle(), 40, ('yellow', 'blue'))
```

`getturtle()` returns the anonymous turtle.

Section 9.2 ~ Exercises

1.   Draw a tic-tac-toe grid

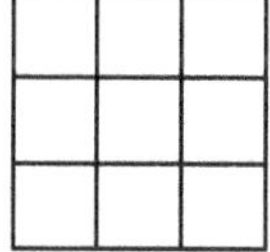

✓

2.   Write a function `draw_polygon(n, a)` that uses a default turtle to draw a
regular polygon with n sides of length a, in the current color. 〈 Hint: if you
have n sides, you need to turn n times and cover the whole 360° angle at the
end, so each turn is $\dfrac{360}{n}$ degrees. 〉

3. (a) Convert the code from Example 2 into a function
         `draw_hexagon(side)` that draws a hexagon with a given side length
         in the current color, centered at the current position. ✓

    (b) Use the function from Part (a) to draw three concentric hexagons of
         different colors. For example:

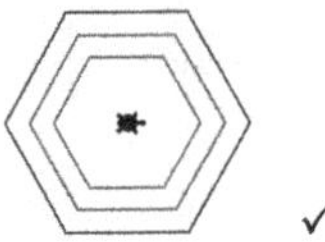

    (c) Add a few lines of code to your solution for Part (b) to make the
         innermost hexagon filled, like this:

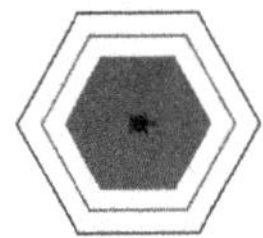

4. Using the `draw_rectangle` function from Example 3 or the
    `draw_polygon` function from Question **2**, draw a path, made of ten
    flagstones, that bends slightly upward. The colors of the flagstones should
    vary, chosen at random among red, blue, dark green, black, and orange. For
    example:

    ⧘ Hint: Recall that the function `choice` from the `random` module returns a
    randomly chosen element of a list. ⧙

5.■ Convert the code from Example 1 into the `draw_triangle` function and
    draw a pyramid of triangles like this:

6.♦ Using the `draw_polygon` function from Question 2, draw a honeycomb filled with "honey" (orange color):

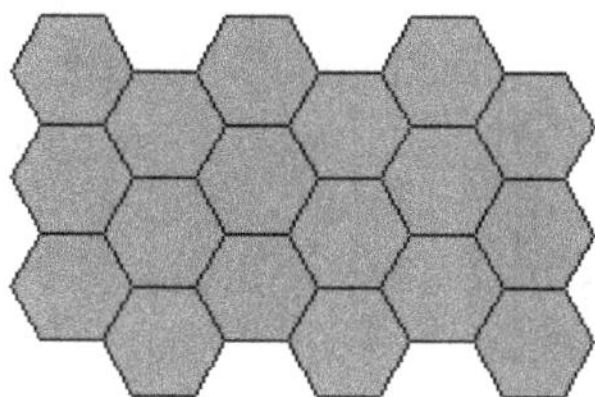

7.♦ Write and test a function to draw a flower with a specified number of petals:

≶ Hint: For a prettier flower, the number of petals should be an odd number. Add 1 if the specified number of petals is even. ≷

9.3   Coordinates and Text

`turtle`'s coordinate system is similar to Cartesian coordinates in math: the *x*-axis is horizontal and points to the right; the *y*-axis is vertical and points <u>up</u> (unlike many other computer graphics packages where the *y*-axis points down). The origin is at the center of the graphics window (Figure 9-3). The default units are pixels. Angles are measured as in math, starting from the positive direction of the *x*-axis and going counterclockwise; the default units for angles are degrees.

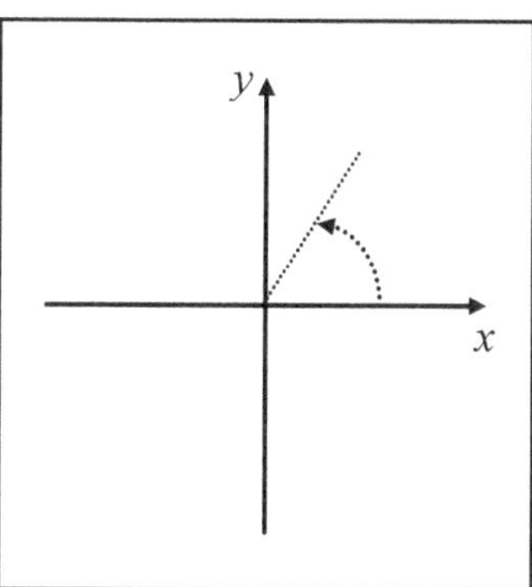

Figure 9-3. Python turtle graphics default coordinates

`turtle` has functions that change the size of the graphics window, the origin of the coordinate system, and the units, but we will stay with the defaults.

So far we have used turtle functions that move and turn the turtle relative to its current position and direction: `forward`, `backward`, `left` and `right`. These commands are easy for a robot to handle, and, in fact, for humans, too. But `turtle` also has several functions that deal with absolute coordinates and angles. These functions are summarized in Table 9-3.

`turtle` also has a function `circle` that draws a circle of a given radius, starting at the current position and direction and going counterclockwise. `circle` also can draw an arc if the `extent` parameter is given. For example:

```
t = Turtle()
t.left(180)
t.circle(80, extent=180)
```

draws a semicircle:

`circle` will actually draw a polygon if the `steps` parameter is specified. For example:

```
>>> from turtle import *
>>> circle(80, steps=5)
>>> hideturtle()
```

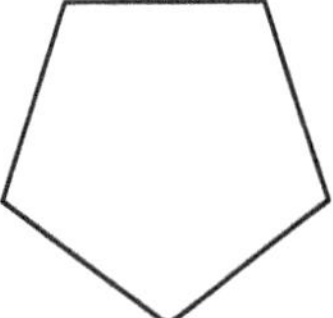

Function	Action
`position()` `pos()`	Return turtle's current *x-y* coordinates (as a tuple).
`xcor()` `ycor()`	Return turtle's current *x*- or *y*-coordinate, respectively.
`distance(x, y)`	Return the distance from the turtle to the point (x, y).
`heading()`	Return turtle's current direction.
`towards(x, y)`	Return the angle from the *x*-axis to the vector (line) from the turtle to the point (x, y).

`setposition(x, y)`
`setpos(x, y)`
`goto(x, y)`
	Move the turtle to the point (x, y); the turtle's direction remains unchanged.

`setx(x)` `sety(y)`	Set the turtle's respective coordinate without changing the other coordinate or direction.
`setheading(to_angle)` `seth((to_angle)`	Change the turtle's direction to `to_angle`.
`home()`	Return the turtle to the origin, pointing east.

Table 9-3. `turtle` functions that use absolute coordinates and angles

The `dot(diameter, c)` function will draw a circular dot of the given diameter, filled with the color `c` and centered at the current turtle's position. `dot(diameter)` draws a dot in the current color.

Example 1

Draw a smiley face:

Solution

```
pensize(2)  # for thicker lines
circle(120)
penup()
setposition(-50, 140)
dot(30)
setposition(50, 140)
dot(30)
setposition(-40, 60)
setheading(-53.13) # the angle in the 3-4-5 triangle is arctan(4/3)
pendown()
pensize(4)
circle(50, extent=2*53.13)
penup()
hideturtle()
```

❖   ❖   ❖

turtle's `write(msg, font=fnt)` function displays the string `msg` in a specified font. For example:

```
t = Turtle()
t.write('Once I was a real turtle.', font=('arial', 20))
```

displays

Once I was a real turtle.

> **The text is displayed in the turtle's current color. The current direction of the turtle does not affect the text and remains unchanged.**

The `font` parameter is a tuple that includes the font name and size, which can be followed by `'bold'`, `'italic'`, and/or `'underline'` in any combination and order. For example:

```
t = Turtle('turtle')
t.color('blue')
t.write('Once I was a real turtle.', font=('Arial', 20, 'bold', 'italic'))
```

Once I was a real turtle.

The available font names are those installed in your operating system, but in a portable program it is advisable to use only common fonts that are available in most systems, such as `'Arial'`, `'Times'`, and `'Courier'`, or just write `None` for the font name to use the default font.

By default, the left end of the baseline of text will be at the current turtle position. An optional parameter, `align='center'` or `align='right'`, will place the center or the right end of the baseline at the current position. The optional parameter, `move=True` will move the turtle to the end of the baseline (and draw if the pen is down). For example:

```
t.write('Once I was a real turtle.',
        font=('times', 20, 'italic'), align='center', move=True)
```

Once I was a real turtle.

If the text string contains `'\n'` characters, `write` will correctly display multiple lines.

Section 9.3 ~ Exercises

1.   Draw a "hamburger" button

2.   The game of *Nim* is, theoretically, played with piles of stones, but it is commonly played with rows of sticks instead. Draw a configuration with three rows of sticks:

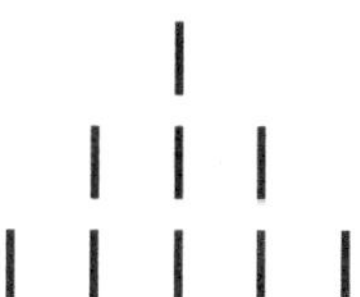

3.   Another, *isomorphic* (mathematically identical) representation of Nim is tokens moving from left to right on a rectangular board. Draw the Nim configuration with three tokens:

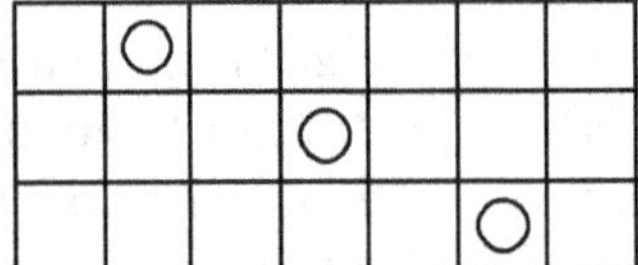

(It is identical to the three rows of sticks in the previous question.) ✓

4.   Draw a snowman:

5.   In the Tower of Hanoi puzzle, you need to transfer a pyramid of disks from one peg to another, using the third peg as a "spare." You can only move one disk at a time, and you may place it only on top of a larger disk or on the base. Draw a two-dimensional sketch of the puzzle with five disks:

6.   Draw a stop sign:

Don't worry about an exact font match — Arial will do for this exercise. ✓

7.   Display the code that draws a hexagon to the right of the hexagon it draws (see Example 2 in the previous section). Use the Courier font for the code.

8.▪   Draw a fairly smooth graph of the parabola $y = x^2$ with a label to its right:

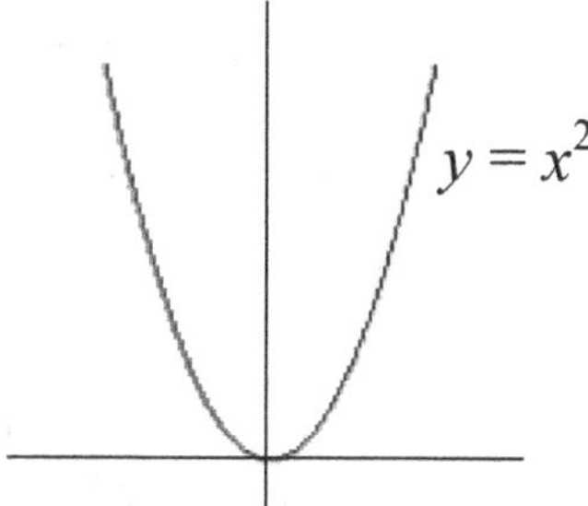

⸎ Hint: generate the segment of the parabola for $-3 \leq x \leq 3$ but scale the graph by a factor of 30 or 40. ⸎ ✓

9.◆   Draw a diagram that illustrates the *golden ratio* (actually taken from the next chapter of this book). Add the equation to the right of the rectangle:

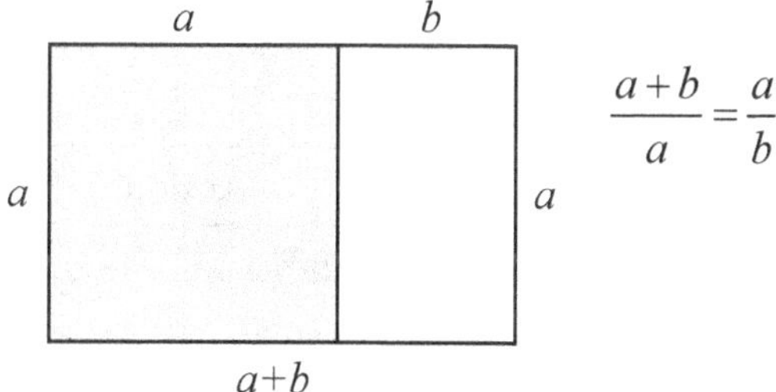

9.4   Colors

In turtle graphics a virtual turtle draws on virtual paper with a virtual pen. No pen exists, of course. What you see on your computer screen is ultimately determined by the contents of the video memory (VRAM) on the *graphics adapter* card or the graphics processor chip. VRAM represents a rectangular array of *pixels* (picture elements). Each pixel has a particular color, which can be represented as a mix of red, green, and blue components, each with its own intensity. A typical graphics adapter uses eight bits to represent each of the red, green, and blue (RGB) values (in the range from 0 to 255). The image on the screen is produced by setting the color of each pixel in VRAM.   The video hardware scans the whole video memory continuously and refreshes the image on the screen.

A graphics processor is what we call a *raster* device: each individual pixel can be set separately from other pixels. (This is different from a *vector* device, such as a plotter, which actually draws lines on paper directly from point *A* to point *B*, with a pen of a particular color.) To draw a red line or a circle on a raster device, you need to set just the right group of pixels to the red color. That's where a graphics package helps: you certainly don't want to program all those functions for setting pixels yourself.

A graphics package has to provide functions for setting colors. Python's `turtle` inherits screen and color handling from the `tkinter` package (Tk interface), which is Python's standard toolkit for GUI (Graphical User Interface) development. `tkinter` uses names assigned to several hundred selected colors. (These names are standard in web app development environments.) You can find some of the named colors with their RGB components in hex and/or decimal form on many web sites, for example, https://trinket.io/docs/colors. A complete list of named colors is available at https://www.tcl.tk/man/tcl8.4/TkCmd/colors.htm.

Table 9-4 summarizes `turtle`'s color functions.

Function	Action

`color(c)`
`color(c1, c2)`
`color()`
	Set pen color and fill color to `c`.
	Set pen color to `c1` and fill color to `c2`.
	Return turtle's current pen color and fill color (each as an RGB tuple or name).

`pencolor(c)` `pencolor()`	Set pen color. Return turtle's current pen color.
`fillcolor(c)` `fillcolor()`	Set fill color. Return turtle's current fill color.
`pensize(w)`	Set the width of strokes to `w`; the default is 1.
`Screen().colormode(256)`	Set RGB tuples scale to 0-255.

Table 9-4. `turtle` color handling functions

The `color` function has two forms: `color(c)` sets the color `c` as both the pen color and the fill color. `color(c1, c2)` sets `c1` as the pen color and `c2` as the fill color.

The parameter `c` can be a literal string that holds the color name. It can also be a tuple of three values, the RGB components, or a literal string that holds `'#'` followed by six hex digits, two for each RGB component. (The RGB values are often expressed in hex because it is convenient to use two hex digits for each component.) For example,

```
>>> color('#25D3A0')
```

sets the red component to 0x25 (decimal 37), the green component to 0xD3 (decimal 211) and the blue component to 0xA0 (decimal 160).

`'#000000'` means black and `'#FFFFFF'` means white.

The `turtle` module uses two modes for representing RGB values in a tuple of three elements. In the first mode, these values are real numbers, scaled to the range from 0 to 1. <u>This is the default mode</u>.

To scale these values back to integers in the usual range, from 0 to 255, use

```
Screen().colormode(255)
```

`Screen()` returns the object `screen` associated with the drawing window. It has functions that control the window size, coordinate units, and other settings. See `turtle` documentation for the list of `screen`'s functions.

There are two other functions that set color — `pencolor(c)` and `fillcolor(c)`; They set the pen color and the fill color, respectively. The supported formats for the parameter `c` in these functions are the same as for `color`.

`color()`, when called without parameters, returns the current pen and fill colors either as their symbolic names (if they were set like that) or as a tuple of RGB values (according to the current `colormode`).

Example 1

What are the RGB values for `'dark salmon'`?

Solution

According to https://www.tcl.tk/man/tcl8.4/TkCmd/colors.htm/ and other web sites, the RGB components for this color are 233, 150, 122.

Example 2

Does the color `'#ff69b4'` have a symbolic name?

Solution

Google "#ff69b4" to find out.

> **Python's `turtle` module and the `screen` object have many more functions — for setting window dimensions, defining stroke width, creating new turtle shapes, getting user input, creating animations, capturing mouse clicks and keyboard events, and so on.**

See https://docs.python.org/3.7/library/turtle.html#module-turtledemo for examples.

Section 9.4 ~ Exercises

1. If each of the three RGB color components is represented in one byte, how many different RGB colors are there? ✓

2. If the screen resolution is 1920 by 1200 and three bytes per pixel are used for color, what is the required size of the video memory?

3. Is the 1920 by 1200 screen aspect ratio close to the golden ratio? ✓

4. What are the RGB values for the color `'maroon'`?

5. Draw a color swatch for the 27 colors formed by combinations of 0, 127 and 255 values for each R, G, and B components:

✓

6.   Draw a rainbow of seven colors on a light gray background:

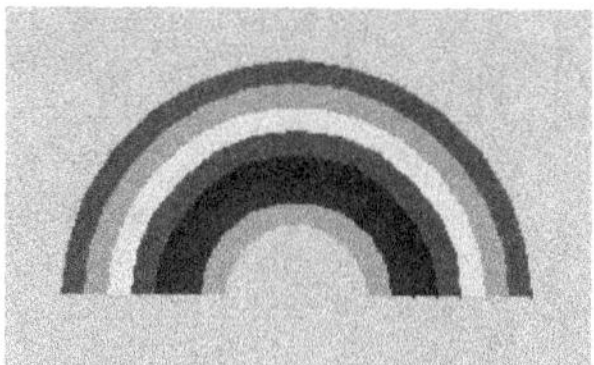

The rainbow colors are red, orange, yellow, green, blue, indigo, and violet.

⸙ Hints:

(a) `Screen().bgcolor(c)` sets the background color of the graphics window.

(b) Draw the rings as overlapping filled semicircles, starting with the largest one and ending with the smallest semicircle in the background color.

7.   Display a smooth gradient from pale pink to bright red:

Now reproduce the famous optical illusion by adding a rectangle of <u>solid</u> mid-range red in the middle:

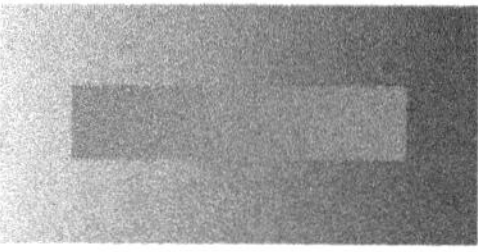

9.5   Review

Terms introduced in this chapter:

> *Logo*
> *Turtle graphics*
> *Module*
> *Virtual*
> *Pixel*
> *Raster device*
> *Graphics adapter*

Some of the Python features introduced in this chapter:

```
from turtle import * # import everything
```

```
t = Turtle() or t = Turtle('turtle')
```

`turtle` **functions:**
```
    speed, forward, backward, left, right, penup, pendown,
    begin_fill, end_fill
    showturtle, hideturtle
    setheading, goto, setx, sety, home
```

`turtle` **color functions:**
```
  color, pencolor, fillcolor, pensize
```

```
Screen().colormode(255)
```

chapter$_{10}$

Sequences and Sums

10.1 Prologue

In math, a *sequence* is an infinite list of values. An element of a sequence is often called a *term*. We will deal only with numeric sequences — those whose terms are real numbers. The terms of a sequence are numbered by integers, starting from 1 or sometimes from 0:

$$a_1, a_2, \ldots, a_n, \ldots$$

or

$$b_0, b_1, \ldots, b_n, \ldots$$

The simplest sequence is the sequence of positive integers themselves: 1, 2, 3, 4, ... Another simple sequence is the sequence of positive odd integers: 1, 3, 5, 7, ...

Sometimes it is not obvious how a sequence is defined, given only its first few terms. For example: 1, 2, 5, 12, 27, ... It turns out we used the formula $a_n = 2^n - n$ to calculate the n-th term. So, it often helps to include the formula for the n-th term in the description of the sequence. For example: 1, 2, 5, ..., $2^n - n$, ... Or we can say simply: "Consider a sequence $\{a_n = 2^n - n\}$." The n-th term formula, such as $a_n = 2^n - n$, is called the *general term* of the sequence.

You can view a sequence of real numbers as a function whose domain is all positive integers (or all non-negative integers) and whose outputs are real numbers: $f(i) = a_i$, for $i = 1, 2, 3, \ldots, n, \ldots$. Like any other function, we can describe a sequence in words or with a formula. Sometimes it's easier to generate a long segment of a sequence with a computer. For example, consider the sequence of all *prime numbers* below 1000: $2, 3, 5, 7, 11 \ldots, 997$. (An integer is called *prime* if it is greater than 1 and is evenly divisible only by 1 and by itself.) In this sequence we can find each term by looking for the smallest number not evenly divisible by any of the preceding terms. Another example: the sequence of digits of π: 3, 1, 4, ...

There is a database of sequences of integers on the Internet, at https://oeis.org/. It is called *The On-Line Encyclopedia of Integer Sequences*. The database contains over 300,000 "interesting" sequences (that is, sequences that came up in one mathematical problem or another). About 10,000 new sequences are added every year.

We say that a sequence *converges* to a certain number (called the *limit* of the sequence) if its terms get closer and closer to that number as n increases. For example, the sequence $\left\{ a_n = \dfrac{1}{2^n} \right\}$ converges to 0. The sequence $1, -1, 1, -1, 1, \ldots$ bounces around zero but does not converge to any limit. The sequence $1, 2, 4, 8, 16, \ldots$ does not converge because its terms get infinitely large as n increases. When a sequence does not converge, we say that it *diverges*.

10.2   Arithmetic and Geometric Sequences

Two types of sequences come up frequently in mathematical problems: the *arithmetic sequence* and the *geometric sequence*.

> **In an arithmetic sequence, the difference between any two consecutive terms is the same.**

In other words, in an arithmetic sequence, $a_n = a_{n-1} + d$, where d is some constant. d is called the *common difference*. Thus an arithmetic sequence has the form $a, a+d, a+2d, \ldots, a+(n-1)d, \ldots$. The general term of an arithmetic sequence can be written as $a_n = a + (n-1)d$. The simplest arithmetic sequence is, again, the sequence of all positive integers.

Example 1

$1, 3, 5, \ldots, 2n-1, \ldots \quad d = 2$.

$5, 15, 25, \ldots, 5 + 10(n-1), \ldots;\ d = 10$.
$3, 0, -3, -6, -9, \ldots;\ d = -3$.

> **In a geometric sequence, the ratio of the next term to the previous term is constant.**

In other words, in a geometric sequence, $a_n = a_{n-1} \cdot r$, so it has the form $a, ar, ar^2, \ldots, ar^{n-1}, \ldots$. The constant r is called the *common ratio*. The general term of a geometric sequence can be written as $a_n = ar^{n-1}$.

Example 2

$1, 2, 4, 8, 16, ...$ and $\dfrac{1}{2}, \dfrac{1}{4}, \dfrac{1}{8}, \dfrac{1}{16}, ...$ are examples of geometric sequences.

The concepts of arithmetic and geometric sequences are related to the concepts of the *arithmetic mean* and *geometric mean*. The arithmetic mean (the average) of two numbers a and b is defined as $\dfrac{a+b}{2}$. The geometric mean of two positive numbers a and b is defined as $\sqrt{ab}$.

Figure 10-1 gives a geometric interpretation of the arithmetic and geometric mean and demonstrates that the geometric mean never exceeds the arithmetic mean: for non-negative a and b $\dfrac{a+b}{2} \geq \sqrt{ab}$, and they are equal only if $a = b$.

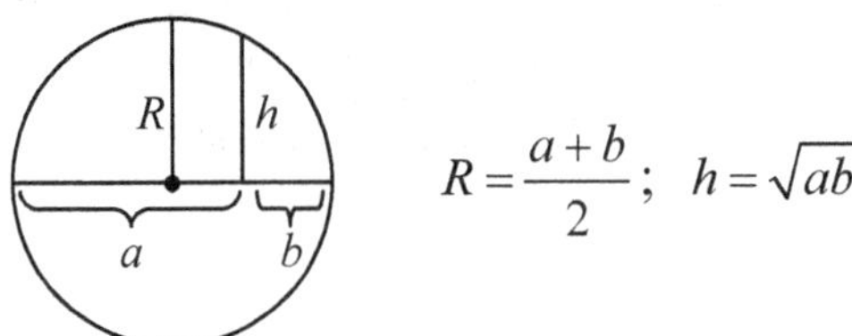

$$R = \frac{a+b}{2}; \quad h = \sqrt{ab}$$

Figure 10-1. Arithmetic mean is greater than or equal to geometric mean

Section 10.2 ~ Exercises

1.   Come up with your own "interesting" sequence, defined in words. State its *n*-th term.

2.   Suggest a formula for the general term of the sequence
$\dfrac{1}{2}, \dfrac{1}{6}, \dfrac{1}{12}, \dfrac{1}{20}, \dfrac{1}{30}, \dfrac{1}{42}, ...$ ✓

3.   Show that if $a_0, a_1, a_2, \ldots$ is an arithmetic sequence, then $a_0, a_3, a_6, a_9, \ldots$ is also an arithmetic sequence.

4.   Suppose the first term of an arithmetic sequence is 3 and the 7th term is 21. Find the 12th term. ✓

5.   What is the common ratio of the geometric sequence $4, 12, 36, \ldots$? What is its general term?

6.   Suppose the first term of a geometric sequence is 1 and the 11th term is 1024. Find the 21st term.

7.   Show that in an arithmetic sequence, each term (except the first) is the arithmetic mean of its left and right neighbors, that is, $a_n = \dfrac{a_{n-1} + a_{n+1}}{2}$. ✓

8.   Show that in a geometric sequence with positive terms, each term (except the first) is the geometric mean of its left and right neighbors, that is,
$$a_n = \sqrt{a_{n-1} a_{n+1}}.$$

9.   Are there any sequences that are arithmetic and geometric at the same time? ✓

10.   What is the limit of the sequence $0, \dfrac{1}{2}, \dfrac{2}{3}, \dfrac{3}{4}, \dfrac{4}{5}, \ldots$? ✓

11.■   Does the sequence $\left\{ a_n = \dfrac{n-2}{n-5} \right\}$ converge? If so, what is the limit?

12.◆   What is the limit of the sequence $\left\{ a_n = \dfrac{n}{\sqrt{n^2 - 1}} \right\}$?

10.3   Sums

In many mathematical situations we are interested in the sum of the first n terms of a sequence.

The sum of an arithmetic sequence

We have already seen in Chapter 4 that $1+2+...+n = \dfrac{n(n+1)}{2}$. There is a simple proof of this fact.

<u>Theorem</u>:

For any positive integer n, $1+2+...+n = \dfrac{n(n+1)}{2}$.

<u>Proof</u>:

Let

$$s(n) = 1 \quad + \quad 2 \quad + \quad ... \quad + \quad (n-1) \quad + \quad n$$

Let's write this sum in reverse order (we know that the result of addition does not depend on the order of the operands):

$$s(n) = n \quad + \quad (n-1) \quad + \quad ... \quad + \quad 2 \quad + \quad 1$$

Now let's add the two lines together (the term above plus the term below):

$$
\begin{array}{cccccccccc}
\boxed{s(n)} = & \boxed{1} & + & \boxed{2} & + & ... & + & \boxed{(n-1)} & + & \boxed{n}\\
\boxed{s(n)} = & \boxed{n} & + & \boxed{(n-1)} & + & ... & + & \boxed{2} & + & \boxed{1}
\end{array}
$$

We get:

$$2\cdot s(n) = \underbrace{n{+}1 \quad + \quad n{+}1 \quad + \quad ... \quad + \quad n{+}1 \quad + \quad n{+}1}_{n \text{ times}}$$

Therefore, $2 \cdot s(n) = n(n+1)$ and $s(n) = \dfrac{n(n+1)}{2}$, Q.E.D. (This is from Latin: Quod Erat Demonstrandum — what we needed to show.)

Ideally, a mathematical proof should follow from the most basic agreed-on assumptions (*postulates*) or from facts that have already been proven earlier (*theorems*). It would be way too tedious, though, to do that for every proof. So mathematicians often skip the earlier steps that depend on established mathematical facts. The above proof is an example of that. Mathematicians know that someone somewhere had worked out the needed steps down to the postulates.

We can find the sum of the first n terms of any arithmetic sequence in a similar way.

> **The sum of the first n terms of any arithmetic sequence is the average of the first and the last terms times the number of terms.**

❖   ❖   ❖

A more interesting example is the sum of the first n terms of the sequence $\left\{a_n = n^3\right\}$, that is, $1^3 + 2^3 + 3^3 + \ldots + n^3$. With a little algebraic technique it is fairly easy to show that this sum is equal to $\left(\dfrac{n(n+1)}{2}\right)^2$. So $1^3 + 2^3 + 3^3 + \ldots + n^3 = \left(1 + 2 + 3 + \ldots + n\right)^2$. It turns out that the sum of the first n cubes is always a perfect square!

❖   ❖   ❖

There is a special "sigma" notation for sums: $a_1 + a_2 + \ldots + a_n$ is often written as $\displaystyle\sum_{i=1}^{n} a_i$. Σ is the letter *sigma* of the Greek alphabet, written in upper case. i is a *variable* — you can use j, k, or any other letter. So we can write the sum of the first ten cubes as $\displaystyle\sum_{k=1}^{10} k^3$. The above identity involving the sum of the first n cubes can be restated as $\displaystyle\sum_{k=1}^{n} k^3 = \left(\sum_{k=1}^{n} k\right)^2$.

The sum of a geometric sequence

Let's start with a simple geometric sequence: $1, 2, 4, ..., 2^{n-1}, ...$. We want to find the sum of the first n terms of this sequence:

$$s_n = 1 + 2 + 4 + ... + 2^{n-1}$$

If we multiply both sides by 2, we get:

$$2s_n = 2 + 4 + 8 + ... + 2^{n-1} + 2^n$$

— a very similar sum, with almost the same terms, except the first term is missing and 2^n is added on the right. If we subtract the first sum from the second, the same terms will cancel out, and we will be left with $2s_n - s_n = s_n = 2^n - 1$.

Question 6 in the exercises asks you to derive a formula for the sum of a general geometric sequence.

Section 10.3 ~ Exercises

1. We have proved that the sum $1 + 2 + ... + n$ is equal to $\dfrac{n(n+1)}{2}$, which is the average of the first and last terms, $\dfrac{1+n}{2}$, times the number of terms, n.

    Make a similar proof to show that the same is true for any arithmetic sequence.

2. Derive the formula for the sum of the first n terms of the arithmetic sequence $\displaystyle\sum_{i=1}^{n}\left[a_1 + (i-1)d\right]$ in a different way, by reducing it to the known sum $1 + 2 + ... + (n-1)$. After simplification, your result should be the same as in Question 1. $\{$ Hint: $\displaystyle\sum_{i=1}^{n} c = nc$; $\displaystyle\sum_{i=1}^{n}\left[(i-1)d\right] = d\sum_{i=1}^{n}(i-1)$. $\}$

3.   Question 3 in Section 4.3 led you to believe that the sum of the first n <u>odd</u> numbers is equal to n^2. Show that this is consistent with the general formula for the sum of an arithmetic sequence.

4.■   Find the sum of the first n terms of the *telescopic* sequence

$$\frac{1}{1\cdot 2}+\frac{1}{2\cdot 3}+\frac{1}{3\cdot 4}+\ldots+\frac{1}{n\cdot(n+1)}.$$ ⌇ Hint: $\dfrac{1}{7\cdot 8}=\dfrac{1}{7}-\dfrac{1}{8}$, for example. ⌇ ✓

5.   Show graphically that $\dfrac{1}{2}+\dfrac{1}{4}+\ldots+\dfrac{1}{2^n}=1-\dfrac{1}{2^n}$. ⌇ Hint: draw a square "pizza", 1 by 1 foot. Cut it in half. Cut one of the pieces in half. Cut one of the new pieces in half. Keep cutting... ⌇

6.■   Derive a formula for $s_n=a+ar+ar^2+\ldots+ar^{n-1}$. Verify the result by applying it to $\dfrac{1}{2}+\dfrac{1}{4}+\ldots+\dfrac{1}{2^n}$. ⌇ Hint: recall what we did with $1+2+4+\ldots+2^n$ and use a similar method. ⌇ ✓

7.■   Find $\displaystyle\sum_{d=1}^{6} d\cdot 10^d$. ✓

8.■   Derive a formula for $\displaystyle\sum_{k=1}^{n}\left(2^k-k\right)$.

9.◆   Derive a formula for $1^2+2^2+\ldots+n^2$. ⌇ Hint:
$$k(k+1)(k+2)-(k-1)k(k+1)=3k(k+1)=3k^2+3k, \text{ so}$$
$$\sum_{k=1}^{n}k^2=\frac{1}{3}\sum_{k=1}^{n}\left(k(k+1)(k+2)-(k-1)k(k+1)\right)-\sum_{k=1}^{n}k.$$ ⌇ ✓

10.4   Infinite Sums

An infinite sum? Can $a_1 + a_2 + a_3 + ... + a_n + ...$ make any sense? Isn't such a "sum" always infinite? When you encounter infinity, it always raises infinitely many questions.

As it turns out, there is a way to give a precise mathematical meaning to an expression $a_1 + a_2 + a_3 + ... + a_n + ...$. Such an expression is called a *series*. The way to approach it is this: consider a sequence of sums $s_1, s_2, ..., s_n, ...$, where

$$s_1 = a_1$$
$$s_2 = a_1 + a_2$$
$$...$$
$$s_n = a_1 + ... + a_n$$
$$...$$

Here s_n is a normal finite sum, for any particular n. It is called a *partial sum* of the series.

> **Given a series $\displaystyle\sum_{i=1}^{n} a_i$, if the sequence of its partial sums $s_n = a_1 + ... + a_n$ converges to a number, that number is called the sum of the series, and we say that the series *converges*.**

You may still be not quite convinced: how can a sequence of partial sums ever converge? Aren't we adding more and more terms to the sum? But it is possible for the sums to converge if we add smaller and smaller amounts.

Example 1

For the geometric series $\dfrac{1}{2} + \dfrac{1}{4} + ... + \dfrac{1}{2^n} + ...$, the n-th partial sum is

$$s_n = \frac{1}{2} + \frac{1}{4} + ... + \frac{1}{2^n} = 1 - \frac{1}{2^n}.$$ As n increases, s_n approaches 1 (because $\dfrac{1}{2^n}$ approaches 0). The partial sums never reach 1, but they get closer and closer to 1 as

n increases, and the sum of the whole series is said to be 1 (see Question 5 in the previous section).

❖   ❖   ❖

A series is often written in "sigma" notation, as follows: $\displaystyle\sum_{i=1}^{\infty} a_i$ or $\displaystyle\sum_{k=0}^{\infty} b_k$. ∞ is the symbol used for infinity. So $\displaystyle\sum_{i=1}^{\infty} \frac{1}{2^n} = 1$.

> **For a series to converge, its terms must be getting smaller and smaller, converging to 0.**

Is the *converse* true? That is, if the terms of a series get smaller and smaller, approaching 0, then does the series necessarily converge? It turns out this is <u>not</u> true.

Example 2

The series $1 + \dfrac{1}{2} + \dfrac{1}{3} + \dfrac{1}{4} + ... + \dfrac{1}{n} + ...$ is called the *harmonic series*. Let us show that the harmonic series diverges. The idea is to split it into non-overlapping finite segments such that the sum of each segment exceeds a fixed number ($\dfrac{1}{2}$, for example). Here is the way to do it:

$$\frac{1}{2} = \frac{1}{2}$$

$$\frac{1}{3} + \frac{1}{4} > \frac{1}{4} + \frac{1}{4} = \frac{2}{4} = \frac{1}{2}$$

$$\frac{1}{5} + \frac{1}{6} + \frac{1}{7} + \frac{1}{8} > \frac{1}{8} + \frac{1}{8} + \frac{1}{8} + \frac{1}{8} = \frac{4}{8} = \frac{1}{2}$$

$$\frac{1}{9} + \frac{1}{10} + ... + \frac{1}{16} > \underbrace{\frac{1}{16} + \frac{1}{16} + ... + \frac{1}{16}}_{8 \text{ times}} = \frac{8}{16} = \frac{1}{2}$$

...

And so on. The segments get longer and longer, but we don't care: we have an infinite supply of terms to play with! So we can find a partial sum of this series that is greater than $\underbrace{\dfrac{1}{2}+\dfrac{1}{2}+...+\dfrac{1}{2}}_{k\ \text{times}}$ for any k, and therefore the harmonic series diverges.

Figure 10-2 shows a peculiar construction based on the harmonic series. The tower rests on one brick and extends to the right as far as we want without any other support! If you (mentally) add the $(n+1)$-th brick at the bottom, with the displacement $\dfrac{1}{n}$, it is not very hard to show that the center of gravity of the top n bricks falls on the right border of the added brick.

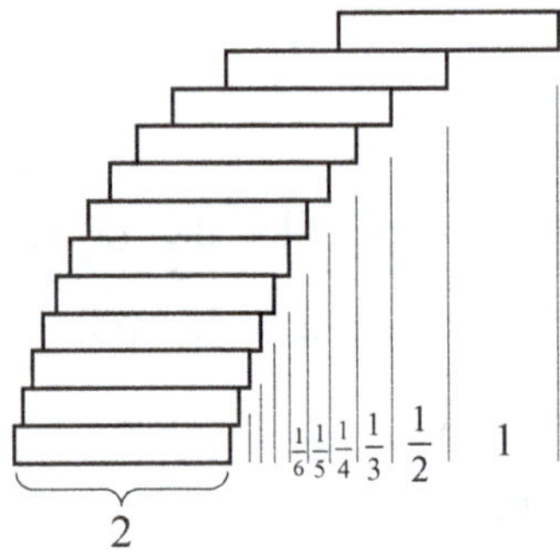

Figure 10-2.   A tower of bricks with one support can extend horizontally as far as we want

Example 3

The series $1+\dfrac{1}{4}+\dfrac{1}{9}+\dfrac{1}{16}+...+\dfrac{1}{n^2}+...$ converges. Moreover, its sum is a surprise: $\dfrac{\pi^2}{6}$. It seems π pops up here out of the blue, but there is a deep mathematical connection.

❖   ❖   ❖

In general, the series is a fascinating topic that reveals many beautiful mathematical facts. Series are studied in Calculus, whose powerful methods help tell whether a particular series converges or not.

Section 10.4 ~ Exercises

1. Find the sum of the *telescopic* series $\dfrac{1}{1\cdot 2}+\dfrac{1}{2\cdot 3}+\dfrac{1}{3\cdot 4}+...+\dfrac{1}{n\cdot(n+1)}+....$

   ≨ Hint: see Question 4 in Section 10.3. ≩

2. Does the series $\displaystyle\sum_{n=1}^{\infty}\dfrac{n+1}{100n}$ converge? Explain. If yes, then what is its sum? ✓

3. If $\{d_0, d_1, d_2,...\}$ is the sequence of digits of π, {3, 1, 4, ...}, does $\displaystyle\sum_{n=0}^{\infty}\dfrac{d_n}{10^n}$ converge? If yes, what is the sum?

4.■ For which values of r does the geometric series $1+r+r^2+...$ converge? ✓

5. Does the series $1+\dfrac{1}{3}+\dfrac{1}{5}+...+\dfrac{1}{2n-1}+...$ converge? Explain your reasoning. ✓

6. Using the built-in function `sum`, write a program the displays n and $6\left(1+\dfrac{1}{4}+\dfrac{1}{9}+\dfrac{1}{16}+...+\dfrac{1}{n^2}\right)$ for $n=1000$ and $n=1000000$. Compare the output to π^2. ≨ Hint: `from math import pi` ≩

7.■ Does the series $\displaystyle\sum_{n=1}^{\infty}\dfrac{1}{n^3}$ converge? Explain your reasoning. ✓

8.♦   The figure below illustrates the process of making a snowflake:

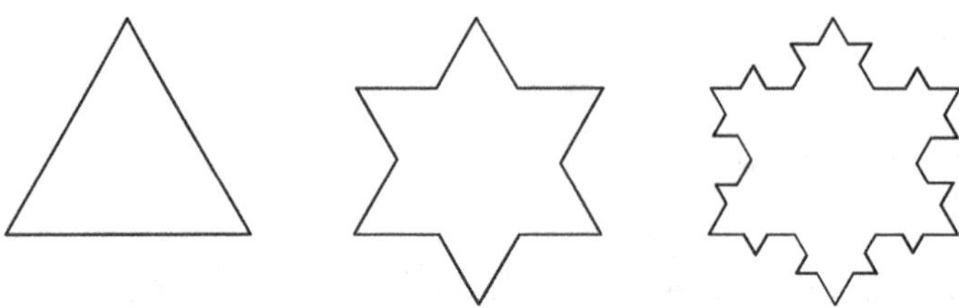

At each step, each straight line segment on the snowflake's perimeter is replaced by —⋀— . If we keep repeating this process infinitely, we get a "curve" known as the *Koch snowflake*. Suppose the perimeter of the first triangle is P and its area is A (where $A = \dfrac{P^2 \sqrt{3}}{36}$). Determine the perimeter and the area of the Koch snowflake after n iterations. Is the perimeter of the Koch snowflake finite? What about the area? ✓

10.5   Fibonacci Numbers

The sequence $\{1, 1, 2, 3, 5, 8, 13, ...\}$ is called *Fibonacci numbers*. In this sequence

$$F_1 = 1; \quad F_2 = 1 \text{ and}$$
$$F_n = F_{n-1} + F_{n-2} \text{ for any } n > 2 .$$

This sequence is named after Leonardo Fibonacci (the name means son of Bonaccio) who lived in the 13th century in the city of Pisa (now in Italy), but the numbers were known in India over 2000 years ago. Fibonacci came upon the numbers while he was modeling the population growth of rabbits under simplistic assumptions: we start with one adult pair; an adult pair produces two baby rabbits each month; a baby rabbit becomes an adult after one month and has its first child when 2 months old; the rabbits never die. In this model the number of adult pairs of rabbits after n months is F_n (Figure 10-3).

Fibonacci numbers pop up in many places in mathematics and computer science, and also in the natural world (see Figure 10-4; and search the Internet for "Fibonacci in nature" for other examples).

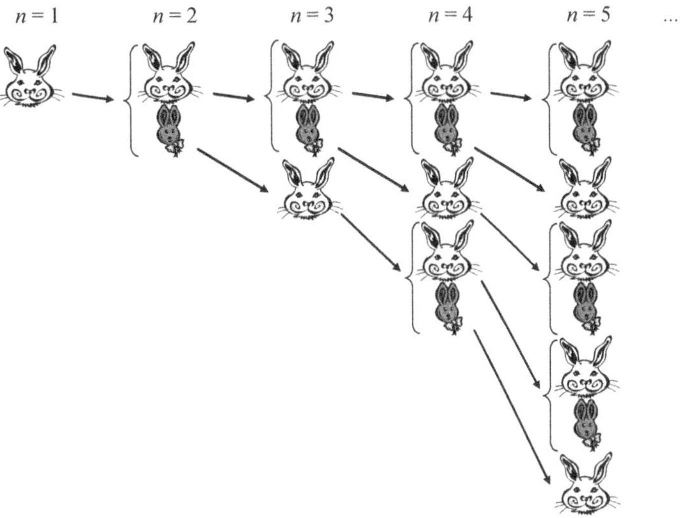

Figure 10-3. Rabbits' population growth in the Fibonacci model

— adult pair   — baby pair

**Figure 10-4.   Broccoli Romanesco. The numbers of clusters in
the spirals form Fibonacci sequences.**

Photograph by Madelaine Zadik, courtesy Botanic Garden of Smith College.

Theorem:

The sum of the first n Fibonacci numbers, $F_1 + F_2 + ... + F_n$ is equal to $F_{n+2} - 1$.

Example:

$1 + 1 + 2 + 3 + 5 = 13 - 1$

Proof:

$$F_1 = F_3 - F_2$$
$$F_2 = F_4 - F_3$$
$$...$$
$$F_n = F_{n+2} - F_{n+1}$$

Adding together corresponding columns we get:
$$F_1 + F_2 + ... + F_n = F_{n+2} - F_2 = F_{n+2} - 1$$

Q.E.D. (Quod Erat Demonstrandum).

Another proof:

The statement of the theorem is true when $n = 2$. Indeed, $1 + 1 = 3 - 1$. Suppose the statement is true for n. Then let us show that it is also true for $(n+1)$.

$$F_1 + F_2 + ... + F_{n+1} = \left(F_1 + F_2 ... + F_n \right) + F_{n+1} = \left(F_{n+2} - 1 \right) + F_{n+1} = \left(F_{n+1} + F_{n+2} \right) - 1 = F_{n+3} - 1$$

— so it is also true for $n + 1$. If the statement of the theorem is true for $n = 2$, then it is true for $n = 3$. If it is true for $n = 3$, then it is true for $n = 4$. And so on — it is true for all n. Q.E.D.

This kind of proof is called proof by *mathematical induction*. The mathematical induction method is explained in Section 16.4.

Fibonacci numbers are defined by a *recurrence relation*. This means that starting at some point in the sequence (in this case starting with F_3), each next term is a function of the previous terms (in this case, $F_n = F_{n-1} + F_{n-2}$). Arithmetic and geometric sequences can be also defined by recurrence relations: $a_n = a_{n-1} + d$ for an arithmetic sequence; $a_n = a_{n-1} \cdot r$ for a geometric sequence. But these two simple sequences can be also described by *closed-form* formulas: $a_n = a_1 + (n-1)d$ and $a_n = a_1 \cdot r^{n-1}$, respectively. It turns out that Fibonacci numbers also have a closed-form formula:

$$F_n = \frac{1}{\sqrt{5}} \left[\left(\frac{1+\sqrt{5}}{2} \right)^n - \left(\frac{1-\sqrt{5}}{2} \right)^n \right]$$

The components of the formula are irrational numbers, but, paradoxically, their combination gives an integer.

How can we come up with this beautiful formula?

Consider the quadratic equation $x^2 = x + 1$. It has two solutions $x = \dfrac{1+\sqrt{5}}{2}$ and $x = \dfrac{1-\sqrt{5}}{2}$. Let's call them φ and $\bar{\varphi}$ (for a reason that we will explain shortly). Notice that $\varphi + \bar{\varphi} = 1$, $\varphi - \bar{\varphi} = \sqrt{5}$, and also $\varphi^2 - \bar{\varphi}^2 = (\varphi + \bar{\varphi})(\varphi - \bar{\varphi}) = \sqrt{5}$.

$\varphi^2 = \varphi + 1 \Rightarrow \varphi^3 = \varphi^2 + \varphi \Rightarrow \varphi^4 = \varphi^3 + \varphi^2 \Rightarrow ...$, so $\varphi, \varphi^2, \varphi^3, ...$ is a peculiar sequence: it is both a geometric sequence and a kind of Fibonacci sequence, only instead of starting with 1, 1 it starts with φ, φ^2. $\bar{\varphi}, \bar{\varphi}^2, \bar{\varphi}^3, ...$ is also a geometric sequence and a kind of Fibonacci sequence; it starts with $\bar{\varphi}, \bar{\varphi}^2$. If we combine these two sequences with any coefficients A and B, we again get a Fibonacci sequence: $A\varphi + B\bar{\varphi}, A\varphi^2 + B\bar{\varphi}^2, ...$. If we choose $A = \dfrac{1}{\sqrt{5}}$ and $B = -\dfrac{1}{\sqrt{5}}$, we will get the standard Fibonacci sequence $\{1, 1, 2, ...\}$. Therefore, $F_n = \dfrac{1}{\sqrt{5}} \left(\varphi^n - \bar{\varphi}^n \right)$.

❖ ❖ ❖

The number $\dfrac{1+\sqrt{5}}{2}=1.61803398875...$ is called the *golden ratio*. It is customary to represent the golden ratio by the Greek letter φ ("Phi").

Why is it "golden"? Consider a rectangle with dimensions $a+b$ and a, such that if we cut an a by a square from it, the remaining smaller rectangle is a similar shape:

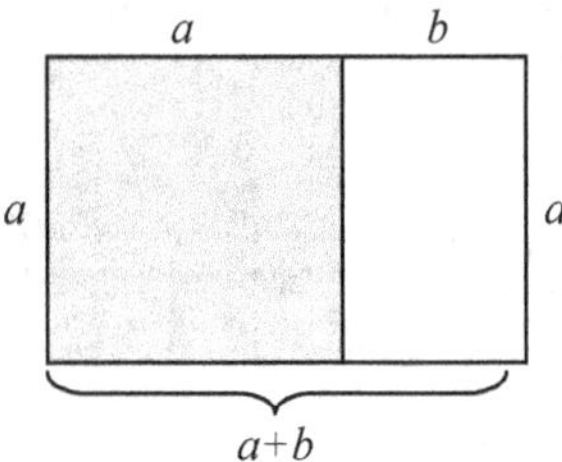

(This shape is considered especially pleasant for the eye.)

"Similar shapes" means that the aspect ratio (the ratio of the longer side to the shorter side) is the same: $\dfrac{a}{b}=\dfrac{a+b}{a} \Rightarrow \dfrac{a}{b}=1+\dfrac{b}{a} \Rightarrow \left(\dfrac{a}{b}\right)^{2}=\dfrac{a}{b}+1$. So $x^{2}=x+1$, where

$x=\dfrac{a}{b}$. This quadratic equation has two solutions: $\varphi=\dfrac{1+\sqrt{5}}{2}=1.612803398875...$

and $\bar{\varphi}=\dfrac{1-\sqrt{5}}{2}=-0.612803398875...$.

The closed-form formula for the n-th Fibonacci number is $F_{n}=\dfrac{1}{\sqrt{5}}\left(\varphi^{n}-\bar{\varphi}^{n}\right)$. For a large n, φ^{n} becomes very large, because $\varphi>1$, and $\bar{\varphi}^{n}$ becomes negligibly small, because $|\bar{\varphi}|<1$. So as n increases F_{n} gets closer and closer to $\dfrac{1}{\sqrt{5}}\varphi^{n}$, and $\dfrac{F_{n}}{F_{n-1}}$ gets closer and closer to φ.

More precisely, $\dfrac{F_{n+1}}{F_n} = \dfrac{\frac{1}{\sqrt5}\left[\varphi^{n+1} - \overline{\varphi}^{n+1}\right]}{\frac{1}{\sqrt5}\left[\varphi^{n} - \overline{\varphi}^{n}\right]} = \dfrac{\varphi - \dfrac{\overline{\varphi}^{n+1}}{\varphi^n}}{1 - \dfrac{\overline{\varphi}^{n}}{\varphi^n}}$. The numerator approaches φ

and the denominator approaches 1, so the limit of $\dfrac{F_{n+1}}{F_n}$ is φ.

> **The limit of the ratio of two consecutive Fibonacci numbers is the golden ratio!**

For example, $\dfrac{13}{8} = 1.625$ and $\dfrac{89}{55} = 1.61818....$

Section 10.5 ~ Exercises

1.   Is F_{1000} odd or even? Explain. ✓

2.   Let $S(n)$ be the number of strings of 0s and 1s that do not contain two 1s in a row. Examine $S(1)$, $S(2)$, $S(3)$, $S(4)$, and $S(5)$, form a *conjecture* (hypothesis) about this sequence, and prove your conjecture. ⸮ Hint: Any string in $S(n)$ ends either with '0' or with '01'. ⸮

3.   Let $T(n)$ be the number of all stripes of length n made of 1-by-1 and 2-by-1 tiles. For example,

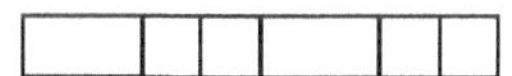

Examine $T(1)$, $T(2)$, $T(3)$, $T(4)$, and $T(5)$, form a conjecture about this sequence, and prove your conjecture. ⸮ Hint: The rightmost tile in any stripe is either a 1-by-1 or a 2-by-1 tile... ⸮

4.   In the diagram below, how many paths along the arrows go from Point A and from Point B to Point C_n?

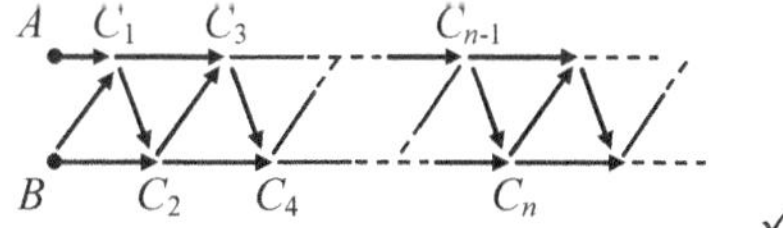

✓

5.   (a)   Fill in the blanks in the following function.

```python
def fibonacci_list(n):
    """ Return a list f of length n+1 in which f[0] is 0
        and f[k] is the k-th Fibonacci number for 1 <= k <= n.
    """
    f = (n+1)*[0]
    f[1] = f[2] = 1

    for ______________________:

        ______________________

    return f
```

Test your function. ✓

(b)   Fill in the blanks in another version of `fibonacci_list`:

```python
def fibonacci_list(n):
    """ Return a list f of length n+1 in which
        f[0] is 0 and f[k] is the k-th Fibonacci number.
    """
    f = [0, 1, 1]

    while ______________________:

        f.append(______________________)

    return f
```

Test this version, too.

(c)▪ Write and test a function `fibonacci(n)` that returns the n-th Fibonacci number and <u>does not use any lists</u>. ⧼ Hint: `f1, f2 = f2, f1+f2`. ⧽ ✓

6.   Experiment with the sums of Fibonacci numbers with odd indices, $F_1 + F_3 + ... + F_{2n-1}$, for different n. Come up with a simple formula for such sums and prove it. ⧼ Hint: `sum(f[k] for k in range(1, 2*n, 2))`. ⧽ ✓

7.   Prove that in the Fibonacci sequence $\{F_n\}$ with $F_1 = 1$, $F_2 = 1$, F_n is the

closest integer to $\dfrac{\varphi^n}{\sqrt{5}}$. ✓

8.   Prove that in the Fibonacci sequence $\{F_n\}$ with $F_1 = 1$, $F_2 = 1$,

$F_{2n+1} = F_n^2 + F_{n+1}^2$ for any $n \geq 1$. First write a program that tests this conjecture for n from 1 to 100 (see Question 5 (a) or (b)). Then provide a mathematical proof. ⸪ Hint: See Question 3 about stripes made of 1-by-1 and 2-by-1 tiles. Take a stripe of length $2n$ and draw a line in the middle. There are two possibilities: either the line crosses a 2-by-1 tile or the line falls between two tiles. ⸫

9.   Prove that in the Fibonacci sequence $\{F_n\}$ with $F_1 = 1$, $F_2 = 1$,

$F_{2n} = F_n^2 + 2F_n F_{n-1}$ for any $n > 1$. ⸪ Hint: Use the result from Question 8. ⸫ ✓

10.   Consider the sequence $\{x_n\}$, where $x_1 = 1$ and $x_n = 1 + \dfrac{1}{x_{n-1}}$ for $n > 1$. Write

a program that prints out the first 100 terms of this sequence. Does this sequence converge? If so, what is its limit? ✓

11.   Derive a closed-form formula for the sequence $\{a_n\}$, where $a_1 = 1$, $a_2 = 5$,

and $a_n = 5a_{n-1} - 6a_{n-2}$ for $n > 2$. What is the limit of $\dfrac{a_{n+1}}{a_n}$?

10.6   Review

Terms and notation introduced in this chapter:

Sequence	*Partial sum*	$\displaystyle\sum_{k=1}^{n} a_k$
Term (of a sequence)	*Converging series*	
Converging sequence	*Diverging series*	$\displaystyle\sum_{k=1}^{\infty} a_k$
Limit of a sequence	*Sum of a series*	
Arithmetic sequence	*Fibonacci numbers*	
Geometric sequence	*Recurrence relation*	
Sigma notation	*Closed-form formula*	
Series	*Golden ratio*	

Some of the Python features introduced in this chapter:

```
f1, f2 = f2, f1+f2
```

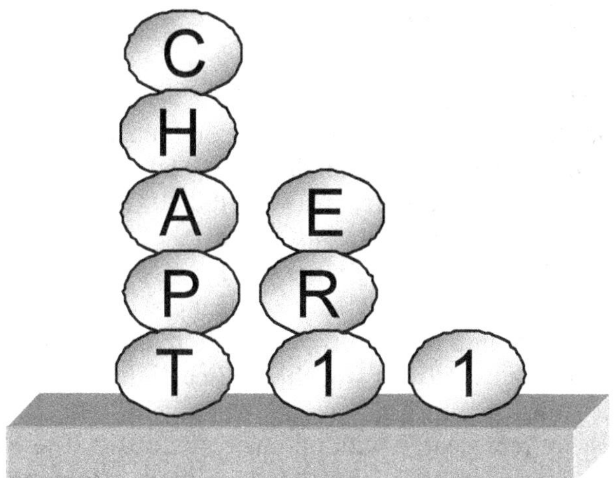

Parity, Invariants, and Finite Strategy Games

11.1 Prologue

Suppose you are buying a box of cereal in a grocery store. The cashier runs the box through the scanner, which reads the UPC (Universal Product Code) barcode from it. Sometimes an error occurs, and the cashier has to scan the same item again. But how does the system know that an error has occurred? If it read the UPC incorrectly, it could potentially charge you for a can of tuna instead of the box of cereal. Fortunately, it turns out that not all 12-digit numbers are valid UPCs. In fact, if you change any one digit in a UPC, you will get an invalid code that does not match any product. This is because not all digits in a UPC carry information: the last digit is a *check digit* that depends on the previous digits. If you change one digit in a UPC, the check digit no longer matches the code. We can say that a UPC has built-in *redundancy*: the information is not represented with optimal efficiency. The check digit is redundant because it can be computed from the other digits. Redundancy allows us to detect and sometimes even correct errors.

When we store or transmit binary data, we can stipulate that the number of 1's in each byte be even (or odd). Only seven bits in each byte would then carry information; the eighth bit would be used as a kind of check digit. It is called the *parity bit*. We will discuss parity and check digits in Section 11.2.

Suppose the data is encoded in bytes with *even parity*. This means the total number of bits set to '1' in each byte is even. If a quantity or property remains constant throughout a process, such a quantity or property is called an *invariant*. The sum of the angles in any triangle is 180 degrees. This is an invariant. If within a `while` loop you add 1 to m and subtract 1 from k, after each iteration the sum $m + k$ remains unchanged. This is an invariant. We will talk about invariants and their role in mathematics and computer science in Section 11.3.

A *combinatorial game* (we call it a *finite strategy game*) is a game with a finite number of possible positions. Two players take turns advancing from one position to the next, according to the rules of the game. Positions never repeat; that is, it is not possible to return to a position visited earlier. Some of the positions are designated as winning positions: the first player to reach a winning position wins. If neither of the players can make a valid move and neither has reached a winning position, then it is a tie. Tic-tac-toe is an example of such a game.

In strategy games of this kind, either the first or the second player always has a winning strategy, or both have a strategy that leads to a tie. In games where a tie is not possible, it is often desirable to describe the winning strategy using some invariant — a property shared by all the winning positions, and by all *safe* positions where the winning player can land. The losing player is always forced to abandon a safe position and move into an *unsafe* one. Nim is an example of such a game. We will discuss the details and see some examples of combinatorial games and their strategies, including Nim, in Section 11.4.

11.2 Parity and Checksums

If you store binary data or send information over a communication line, some errors might occur. A simple method for detecting errors is based on *parity*. Usually the data is divided into relatively short sequences of bits, called *codewords* or *packets*.

> **The term *parity* refers to the evenness or oddness of the number of bits set to '1' in a packet. If this number is even, we say that the packet has *even parity*; otherwise we say it has *odd parity*.**

When a chunk of data is transmitted, the transmitter computes the parity of data and adds one bit (called the *parity bit*), so that the total parity of the packet including the parity bit is even. Then, if the receiver gets a packet with odd parity, it reports an error. (Or the transmitter and receiver may "agree" to use odd parity instead of even parity.) When parity is used, all data packets usually have the same length. For example, seven data bits plus one parity bit, or 31 data bits plus one parity bit.

Example 1

The following sequence of seven-bit codes encodes the word "parity" in ASCII:

```
1110000 1100001 1110010 1101001 1110100 1111001
```

We want to add a parity bit at the end of each code so that it gets even parity. What are the resulting eight-bit packets?

Solution

```
11100001 11000011 11100100 11010010 11101000 11110011
```

❖   ❖   ❖

It can happen, of course, that two errors occur in the same packet — two bits are flipped from 0 to 1 or from 1 to 0. Then the parity of the packet remains unchanged, and the error goes undetected. The parity method relies on the assumption that the likelihood of two errors in the same packet is really low. If errors are frequent, then even a small number of bits requires a parity bit for error checking. The more reliable a communication channel or a storage system is, the longer the data packets that can be used.

If we swap two consecutive bits in a data packet, its parity does not change. Luckily such *transposition errors* are very rare when the packet is generated by a computer or another device. Not so with us humans. When we type words or numbers, transposition errors are common. So parity-type error detection does not work well when humans are involved. For example, when a cashier gives up on the scanner that cannot read the UPC from a crumpled bag of chips, he enters it manually. The cashier may mistype a digit or transpose two digits. There needs to be a mechanism that detects such errors. Such a mechanism uses *checksums* and *check digits*.

Example 2

Driver's licenses on the island of Azkaban have six digits. The sixth digit is the check digit. It is calculated as follows: we add up the first five digits, take the resulting sum modulo 10 (the remainder when the sum is divided by 10), and subtract that number from 10. If we get 10, we replace it by 0. For example, if the first five digits of a driver's license are 95873, the check digit is

$$10 - ((9 + 5 + 8 + 7 + 3) \bmod 10) = 10 - (32 \bmod 10) = 10 - 2 = 8$$

So 958738 is a valid driver's license number on Azkaban. Note that if we add all six digits and take the result modulo 10, we get 0. Does this system detect all single-digit substitution errors? All transposition errors?

Solution

This system detects all single-digit substitution errors, because if you change one digit, the sum of the digits modulo 10 is no longer 0. However, the sum does not depend on the order of digits, so transposition errors are not detected.

To detect both substitution and transposition errors we need a slightly more elaborate algorithm for calculating the check digit. We can multiply some of the digits by certain coefficients before adding them to the sum. Several real-world checksum algorithms are described in the exercises.

Section 11.2 ~ Exercises

1. Which of the following bit packets have even parity? Odd parity?
(a) `01100010` (b) `11010111` (c) `10110001`.

2. How many bytes (all possible eight-bit combinations of 0s and 1s) have even parity? Odd parity? ✓

3. Write and test a Python function that takes a string of binary digits and returns its parity (as an integer, 0 for even, 1 for odd). ⸖ Hint: a string has a method `count`. ⸖

4. Consider the following table of binary digits:

```
0 1 1 0 1 1
1 0 0 0 1 0
1 0 1 0 0 1
0 1 0 1 0 0
```

It was supposed to have even parity in all rows and all columns, but an error occurred and one bit got flipped. Which one? ✓

5.♦ Write and test a Python function `correct_error(t)`, which takes a table, such as described in Question 4 (but not necessarily 4 by 6), with a possible single-bit error, checks whether it has an error, and, if so, corrects it. The table `t` is represented as a list of its rows; each row is represented as a string of 0s and 1s (all the strings have the same length). For example, the table from Question 4 would be represented as

```
['011011', '100010', '101001', '010100']
```

⸖ Hints: find the row with odd parity; if found, find the column with odd parity and "flip" 0 to 1 or 1 to 0 in the string that corresponds to the odd-parity row in the position that corresponds to the odd-parity column. ⸖ ✓

6.■ How many 4 by 6 tables of binary digits have even parity in all rows and all columns? Odd parity? ✓

7.   A UPC has 12 decimal digits. If `s` is a string that represents a UPC, the checksum is calculated as follows:

```
3*s[0] + s[1] + 3*s[2] + s[3] + .... + s[11]
```

(every other digit, starting from the first, is multiplied by 3). In a valid UPC, the checksum must be evenly divisible by 10. For example, `072043000187` is a valid UPC, because
$3 \cdot 0 + 7 + 3 \cdot 2 + 0 + 3 \cdot 4 + 3 + 1 + 3 \cdot 0 + 0 + 3 \cdot 0 + 1 + 3 \cdot 8 + 7 = 60$. Write and test a Python function `is_valid_UPC(s)` that takes a string of 12 digits and returns `True` if it represents a valid UPC; otherwise it returns `False`. ✓

8.   (a) Does the checksum method for UPCs described in Question 7 detect all single-digit substitution errors? ✓

(b)■ In the UPC checksum algorithm described in Question 7, the odd-placed digits are multiplied by 3. Why 3 and not 2? ✓

(c)◆ Does the checksum method for UPCs described in Question 7 detect all transposition errors? ✓

9.■   Credit card numbers have 15 or 16 digits. The checksum is calculated as follows. Each digit in an odd position (first, third, etc.) <u>counting from the end</u> is added to the checksum. Each digit in an even position (second, fourth, etc.) is multiplied by 2. If the result is a one-digit number, it is added to the checksum; if it is a two-digit number, the sum of its two digits is added to the checksum. The checksum must be divisible by 10. Examples:

> `4111 1715 1913 1178` is valid: starting from the right
> $8 + 5 + 1 + 2 + 3 + 2 + 9 + 2 + 5 + 2 + 7 + 2 + 1 + 2 + 1 + 8 = 60$.

(The second addend is 5 because the second digit from the right is 7, $2 \cdot 7 = 14$, and $1 + 4 = 5$.)

> `379 6262 1994 1007` is valid: starting from the right
> $7 + 0 + 0 + 2 + 4 + 9 + 9 + 2 + 2 + 3 + 2 + 3 + 9 + 5 + 3 = 60$.

Write and test a Python function that checks whether a given string of digits represents a valid credit card number. Come up with a few other valid numbers for testing (being careful not to use your own or your family's real credit cards) as well as a few invalid numbers.

10.■   The book industry uses ISBNs (International Standard Book Numbers) to identify books. In 2007 the industry switched from 10-digit ISBNs to 13-digit ISBNs. The last digit in ISBN-10 and in ISBN-13 is the check digit. But the algorithms for calculating the check digit are different for ISBN-10 and ISBN-13.

In ISBN-13, the check digit is calculated the way similar to UPC, only the first digit from left is added to the sum, the second digit from left is multiplied by 3, and so on.

ISBN-13 is obtained from ISBN-10 by appending '978' at the beginning and recalculating the check digit.

Write a Python function `isbn_13_check_digit` that calculates and returns the ISBN-13 check digit (a single character) for a given string of 12 digits, and another function, `isbn_10_to_13`, that converts ISBN-10 (a string of 10 digits) to ISBN-13 and returns a string of 13 digits. Test your functions thoroughly. Use these test data, for example:

ISBN-10	ISBN-13
0982477503	9780982477502
0982477511	9780982477519
098247752X	9780982477526
0982477538	9780982477533
0982477546	9780982477540
0982477554	9780982477557
0982477562	9780982477564
0982477570	9780982477571
0982477589	9780982477588
0982477597	9780982477595

11.■   Question 10 asks you to write two functions that help convert ISBN-10 to ISBN-13. Now write two similar functions to convert ISBN-13 to ISBN-10. In ISBN-10, the check digit is calculated as follows. The first digit is multiplied by 1, the second by 2, the third by 3, and so on; the ninth digit is multiplied by 9. The products are added together and the result is divided by 11. The remainder is used as the check digit. If the remainder is 10, the letter 'X' is used as the check "digit."

11.3 Invariants

If I have several lollipops and give you some, and you give some to Candy, and she gives some back to me and you, the total number of lollipops among the three of us remains the same (as long as we don't eat any) — it is an *invariant*. If a particle moves along a circle, its distance from the center of the circle remains constant — it is an invariant. If a bishop moves on the chessboard along a north-east to south-west diagonal, the sum of the bishop's row and column positions, counting from the upper-left corner, remains the same — it is an invariant. The concept of an invariant is useful in physics, in mathematics, and in computer science.

Example 1

If you toss a small rock in the air, its energy consists of the kinetic energy $\dfrac{mv^2}{2}$ and the potential energy mgh, where m is the mass of the rock, v is its speed, h is the height above ground, and $g = 9.8$ m/sec^2 is the acceleration due to gravity. How far up above the ground will the rock fly if it was tossed straight up from ground level with an initial velocity of 20 m/sec?

Solution

$\dfrac{mv^2}{2} + mgh$ is an invariant, it remains constant. So $\dfrac{mv_0^2}{2} + mgh_0 = \dfrac{mv_1^2}{2} + mgh_1$

At the beginning, $h_0 = 0$, $v_0 = 20$ m/sec. At the top, $v_1 = 0$. Comparing the total energy at the ground and at the top, we get $\dfrac{mv_0^2}{2} = mgh_1 \implies h_1 = \dfrac{v_0^2}{2g} \approx 20.4$ meters.

❖   ❖   ❖

In mathematics, invariants are ubiquitous. One of the applications is in strategy games.

Example 2

A room contains a round table and three bags of coins: one with quarters, one with dimes, and one with nickels. Two players take turns picking one coin from any bag and placing it anywhere on the table without overlapping any other coins. There are enough coins in each bag to cover the entire table. The player who places the last coin, leaving no space for more coins, wins. Does the first or a second player have a winning strategy?

Solution

In this game, there is an infinite number of possible configurations of coins on the table. But the game always ends after several moves, because only a finite number of coins fit on the table. There are no ties.

One approach to finding a strategy in games of this type is to come up with a clever invariant condition, a kind of "balance," which the winner can maintain, always moving into a safe position where the condition is satisfied and forcing the opponent to abandon this safe position, to lose "balance." In this particular game, symmetry with respect to the center of the table comes to mind as a useful invariant. The first player can establish symmetry by placing the first coin at the center of the table. On subsequent moves, the first player always picks the same size coin as the opponent and places it symmetrically to the one just placed by the second player. The first player always maintains the symmetry of the configuration with respect to the center of the table, an invariant. The second player is always forced to break the symmetry. At the end, the first player places a coin in the last remaining spot.

In computer science, the concept of invariant applies to loops. A condition that is related to the purpose of the computation and holds true before the loop and after each iteration through the loop is called a *loop invariant*. Loop invariants are useful for formally proving correctness of the code.

Example 3

The code below has one loop:

```python
def add_numbers(n):
    '''Return 1 + 2 + ... + n.'''
    s = 0
    k = 1
    while k <= n:
        s += k # add k to s
        k += 1 # increment k by 1
    return s
```

Find a loop invariant for that loop.

Solution

The purpose of the loop is to calculate $1+2+...+n$. Before the loop, $s=0$ and $k=1$. After the last iteration through the loop, $k=n+1$ and $s=1+2+...+n$. The loop invariant here is $s=1+2+...+(k-1)$.

Section 11.3 ~ Exercises

1. A domino covers exactly two squares on a chessboard, so it is possible to cover the board with 32 dominos. Now suppose we cut out the two white squares at opposite corners of the board. Try to cover the remaining 62 squares with 31 dominos. Is it possible? If not, explain why not. ✓

2. In chess a knight moves by two squares in one direction, then by one square in a perpendicular direction:

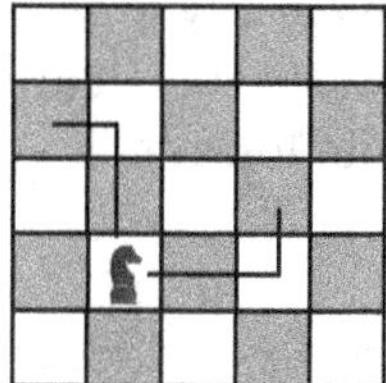

Is it possible for a knight to visit each square on a 8-by-8 board exactly once and return to the starting position? If yes, show an example; if not, explain why not. Is it possible on a 7-by-7 "chessboard"?

3. Consider a rectangle $AOBC$ on the coordinate plane such that O is the origin, AO is on the *x*-axis, OB is on the *y*-axis, and C is in the first quadrant. Describe the locus of points (that is, the set of all points) C such that the perimeter of the rectangle is equal to p, a constant. ✓

4. Consider a rectangle $AOBC$ in the first quadrant on the coordinate plane such that O is the origin and C moves along a branch of the hyperbola $y = \dfrac{1}{x}$.

Describe an invariant property of the rectangle (beyond the obvious fact that O stays at the origin).

5. Demonstrate geometrically that among all the rectangles with a given area, the square has the smallest perimeter. ⸨ Hint: see Questions 3 and 4. ⸩ ✓

6. Several pluses and minuses are written in a line, for example: $+ - - - + + - - + + + -$. If the first two symbols at the beginning of the line are the same, you add a plus at the end; if they are different, you add a minus. Then you erase the first two symbols. The operation is repeated until only one plus or minus remains. Is the remaining symbol always the same, regardless of whether you have proceeded from left to right or from right to left? ✓

7.■   Erin and Owen share a computer. They want to make a schedule for the exclusive use of the computer from noon to midnight, and they have decided to turn this into a game. On each move, a player can reserve a contiguous block of available time, up to two hours, starting at any time. The players take turns making reservations. They have tossed a coin, and Erin goes first. Does Owen have a strategy that would allow him to get at least as much total time as Erin, no matter what she does?

8.◆   The table below lists the numbers of vertices, edges, and faces in four polyhedrons:

	Vertices	Edges	Faces
Tetrahedron	4	6	4
Cube	8	12	6
Triangular prism	6	9	5
Icosahedron	12	30	20

Describe an invariant that connects the number of vertices V, the number of edges E, and the number of faces F in a polyhedron. Show that this is indeed an invariant for all polyhedrons. In fact, the edges and faces do not have to be "straight:" the same invariant remains as long as the edges and faces do not intersect.

9.   Identify a loop invariant for the while loop in

```
k = 0
p = 1
while k < n:
    p *= x
    k += 1
```

✓

11.4 Finite Strategy Games

Suppose a game has a finite number of possible positions. Two players take turns advancing from one position to the next according to the rules of the game. Positions never repeat: a position that occurred once can never happen again. So sooner or later, when a *terminal* position is reached, the game ends. Depending on the rules, the player who made the last move wins or loses, while some of the terminal positions may be designated as ties. Alternatively, the players may collect some points along the way, and the player with the higher score wins. Games of this type are called *combinatorial games*. (We call them *finite strategy games* for clarity.) In some games, such as Nim, reaching a terminal position (taking the last stone) signifies a win, and there are no ties. We will discuss real Nim a little later; first, let us consider a very simple version.

Example 1

There is a pile of N stones. Two players take turns making moves. On each move, a player can take one, two, or three stones from the pile. The player who takes the last stone wins. Let's call this game Take-1-2-3. Does the first or the second player have a winning strategy? What is that strategy?

Solution

This game is equivalent to (a mathematician might say *isomorphic*, that is, "has the same form" as) the game where the players advance a token along a linear board from left to right; on each move a player can advance the token by one, two, or three squares:

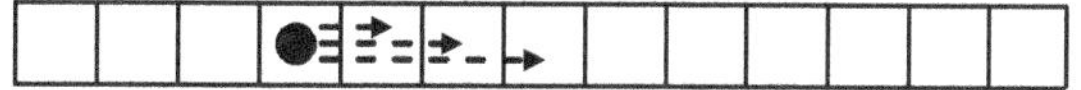

The length of the board is $N+1$. The position on the board corresponds to the number of stones left in the pile; the leftmost square corresponds to N stones, the rightmost square corresponds to 0 stones in the pile. (To reason about a strategy game, it is often convenient to choose a model that is isomorphic to the game but easier to visualize and work with.)

To solve this game, we will use a work-backwards method. Let us first mark the winning position with a plus sign:

This position is safe: that's where you want to end up. Now let us find all the positions from which the plus position can be reached. These are unsafe for you: if you land there, your opponent can jump to a plus. Let us mark each of these unsafe positions with a minus sign:

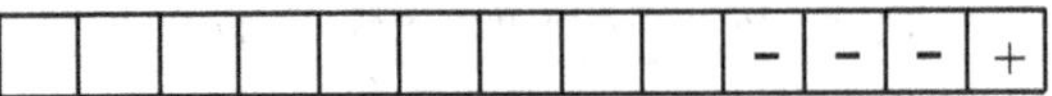

Now there is one position from which your opponent can move only to a minus (unsafe position). This position is safe for you, so let's mark it with a plus:

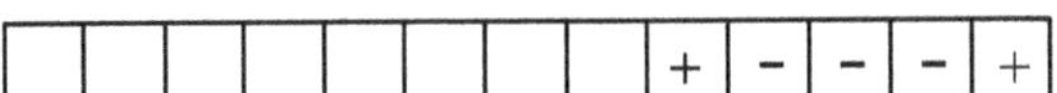

Again let us find all the positions from where one can reach any plus position and mark them with minuses. Continuing this process, we will eventually mark up the whole board:

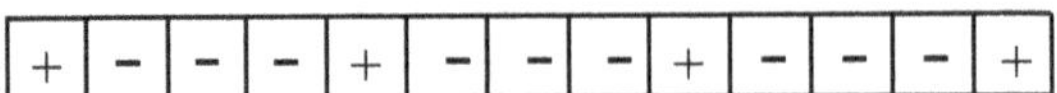

If the starting position is safe (marked with a plus), the first player is forced to abandon it, and the second player can win, moving to a safe position on every turn. In Take-1-2-3 this happens when the initial number of stones $N = 4k$ (N is evenly divisible by four). If the starting position is unsafe (marked with a minus), then the first player can move to a safe position right away, and eventually win. In Take-1-2-3 this happens when N is not evenly divisible by four.

We can represent all the positions in a game as points and each legal move as an arrow from one position to the next. As explained later, in Chapter 17, such a structure is called a *directed graph*. The fact that game positions never repeat means that the graph is *acyclic*; that is, it has no circular paths. Representing games as graphs is useful for developing the mathematical theory of combinatorial games. A lot is known about graphs! But it is too cumbersome for analyzing a particular game, even one as simple as Take-1-2-3. It is better instead to look for a simple enough property or formula that describes all the safe positions.

There is a simple formula for the safe positions in the Take-1-2-3 game: it is safe to move to a position where the number of stones remaining in the pile is evenly divisible by 4. So there is no need to remember the chart of all safe positions; it is much easier to use the formula.

Not every game, however, has a simple formula. Let us consider a more interesting game, for which a simple formula for safe positions hasn't been found yet.

Example 2

In the game of Chomp, the board represents a rectangular grid (like a bar of chocolate). Players take turns taking "bites" from the grid. A "bite" consists of one square plus all the squares to the right and/or below it. For example:

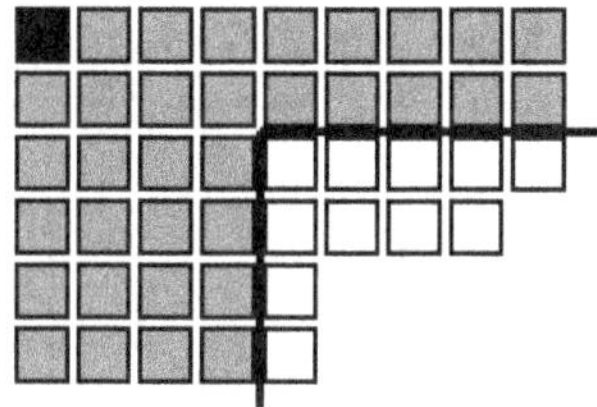

The square in the upper left corner is "poison"; the player who is forced to "eat" the poison loses the game.

It is possible, of course, to apply to Chomp the work-backwards method described in Example 1 and come up with a list of all safe positions (see Question 7 in the exercises). Such analysis, however, would be very tedious for larger boards if we tried doing it by hand. It is better to program a computer to do that. No one, so far, has been able to come up with a compact property or formula that describes all safe positions in Chomp, except for two special board sizes: n-by-2 and n-by-n (see Question 6 in the exercises).

One peculiar thing about Chomp is that we can prove, even without knowing anything about any specific strategy, that the first player can always win (as long as the board is larger than 1 by 1). Here is a proof by contradiction. Suppose it is the second player who has a winning strategy. So the second player has a winning response to each of the first player's moves. If the first player "bites off" just one square, at the lower right corner, in the first move, the second player has a winning move in response. But the first player could have made that winning move first! This proof is based on the argument called *strategy stealing*: in this game the second player can't have a winning strategy because the first player would "steal" it.

Our last example in this section is the real game of Nim (*nimm* is German for "take"). In Nim there are several piles of stones. On each move, a player must take at least one stone but can take any number of stones from one pile. The player who takes the last stone wins. (There is another version of Nim in which the player who takes the last stone loses. The winning strategy in this take-last-and-lose Nim is similar to our Nim.)

Nim is sometimes presented as an arrangement of cards or tokens or matches in several rows. For example:

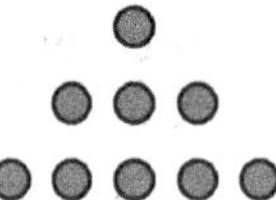

Each row represents a pile of stones. Another isomorphic model for Nim is several tokens on a rectangular board moving in the same direction. For example:

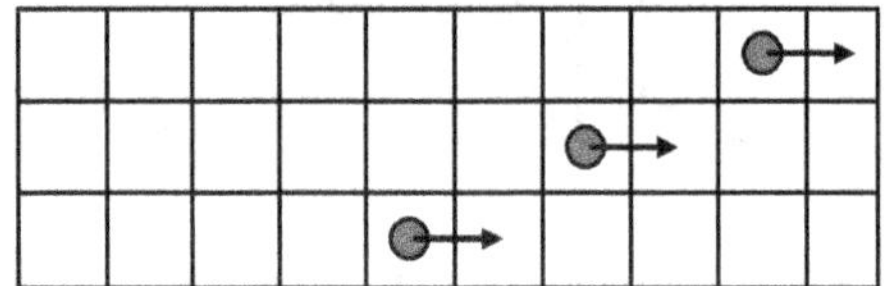

Each token's distance from the rightmost square represents the number of stones in the corresponding pile.

It is possible to use the work-backwards method to determine a strategy for Nim. This is not very interesting, though, and quite tedious when the numbers are large. Luckily, Nim has a very elegant description of the winning strategy based on an interesting property of its safe positions.

Suppose N_1, N_2, ..., N_k are the numbers of stones left in the piles. The idea is for the winning player to always maintain some kind of "balance" among these numbers. The losing player is forced to change one of the numbers and break the balance; then the winning player restores the balance again. But what kind of balance? Perhaps some kind of checksum might work. When one of the numbers changes, the checksum will change, too. The problem is, the winning player must be able to restore the checksum by <u>reducing</u> one of the numbers. Conventional checksum algorithms don't work like that. We need something more ingenious.

Let us arrange the binary representations of N_1, N_2, ..., N_k one under another, with the rightmost digits aligned. For example, if the numbers are 1, 3, 5, and 7, we get

$$
\begin{array}{ll}
N_1 & 1 \\
N_2 & 11 \\
N_3 & 101 \\
N_4 & 111
\end{array}
$$

Now consider the parity of each column of binary digits — whether the number of 1s in the column is even or odd. In the final winning position, all the numbers are zeros, so the parity of all the columns is even. Let us declare "safe" all positions with this property: the parity of all columns is even. All other positions are "unsafe."

The parities of the columns can be considered as binary digits of a kind of "checksum." Calculating this "checksum" is equivalent to performing bitwise addition, or — same thing — the bitwise XOR (exclusive OR) operation on the numbers (see Section 8.3). This "checksum" is called the *Nim sum*. In safe positions the Nim sum must be 0. For example:

$$
\begin{array}{llcll}
& \textit{Safe:} & & & \textit{Unsafe:} \\
N_1 & 1 & & N_1 & 1 \\
N_2 & 11 & & N_2 & 1 \\
N_3 & 101 & & N_3 & 101 \\
N_4 & 111 & & N_4 & 111 \\
& === & & & === \\
\textit{Nim sum} & 000 & & & 010
\end{array}
$$

It is very easy to calculate the Nim sum on a computer (see Question 10 in the exercises).

From a safe position you are forced to move to an unsafe one. Indeed, when a player takes one or several stones from the j-th pile, N_j changes, so at least one of its binary digits changes. The parity of the column that holds that digit will change, too. For example, in the 1-3-5-7 configuration, the numbers of bits in the three columns are 2, 2, and 4 — all even, so this is a safe position. The first player is forced to abandon it on the first move, so the second player can win.

Is it true, though, that from any unsafe position you can always move to a safe one? In other words, is it true that the even-parity-of-all-columns property can always be restored by <u>reducing</u> one of the numbers? The answer is yes, and here is why. Suppose some of the columns have odd parity. Let's take the leftmost of them. Since its parity is odd, it must have at least one bit in it set to 1. Let's take the row that contains that bit and flip (from 1 to 0 or from 0 to 1) all the bits in that row that are in odd-parity columns. The even parity of all columns will be restored. The new number represented by the row will be smaller than the original number, because the leftmost flipped bit (that is, the leftmost binary digit flipped) has changed from 1 to 0.

For example, suppose that in the 1-3-5-7 configuration, the first player removes the entire second pile. The numbers become 1, 0, 5, 7:

N_1	1
N_2	0
N_3	101
N_4	111
	===
Nim sum	011

The parity of the second and third columns becomes odd. The second column is the leftmost among them. The bit in the fourth row is set to 1. Let's take that row and flip the bits in it that are in the odd-parity columns. (This is equivalent to XOR-ing that row with the Nim sum.) We get:

N_1	1
N_2	0
N_3	101
N_4	100
	===
Nim sum	000

The even parity of all columns is restored: the Nim sum is 0 again. N_4 becomes 4. So to restore the balance and return to a safe position from the 1-0-5-7 position, the second player should take from the fourth pile 3 stones out of 7, leaving 4 stones.

Nim is important, because there are many more general versions of it (see, for example, Question 11 in the exercises), and many games are isomorphic to a version of Nim.

Section 11.4 ~ Exercises

1.   What property describes the safe positions in the Take-1-2-3-4 game, in which you have one pile of stones and are allowed to take 1, 2, 3, or 4, stones on each move? ✓

2.■   Write a Python program that plays Take-1-2-3 with the user. Start with a reasonable random number of stones in the pile, say, 8 to 12. Let the human player go first. Prompt the human player to make a move; accept only valid input (1, 2, or 3, but not exceeding the total remaining number of stones in the pile); explain invalid input attempts until the player makes a valid move. When it is the computer's turn and it is in an unsafe position, it should take the number of stones to move to a safe position. If the computer is in a safe position, it should take randomly 1, 2, or 3 stones. You might want to implement the "human's" move and the computer's move in separate functions that return the number of stones taken. Declare a winner when there are no stones left in the pile.

3.   In this game, two players take turns moving a token on an 8-by-8 board. At the beginning, the token is placed in the lower left corner. On each move, the token can be moved by one square up or to the right or diagonally up and to the right:

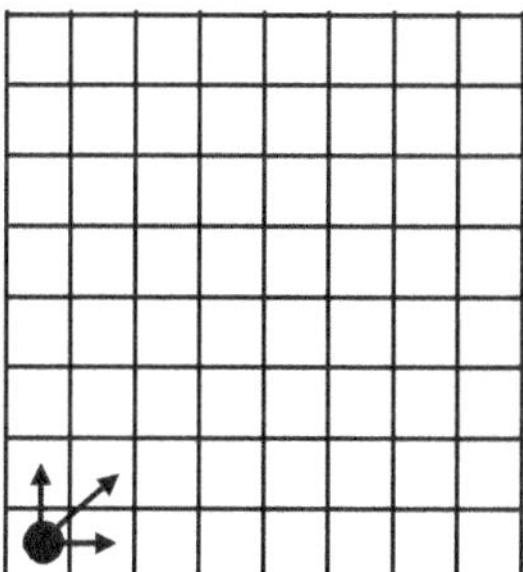

The player who reaches the upper right corner first wins. Find all safe positions in this game. Can the first player always win? ✓

4.   Describe an isomorphic version of the game from Question 3, as a game that uses piles of stones instead of the board. Describe the safe positions in your version. ✓

5. Suppose the game from Question 3 has been modified: now the field has a poisonous swamp, like this:

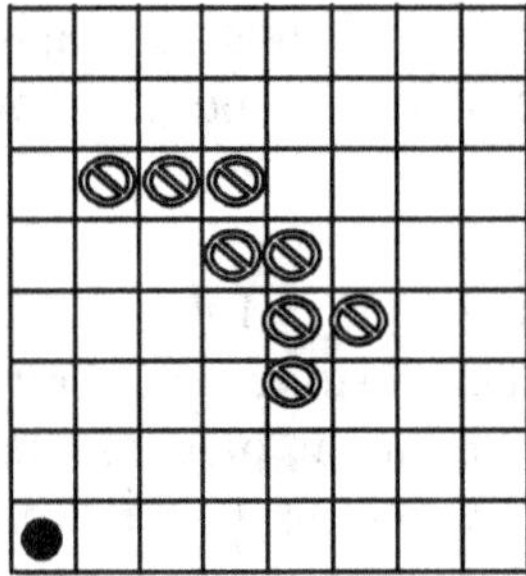

The player who has no valid move or is forced into the swamp loses. Find all safe positions on the above board and show that the first player can win. If the first player moves diagonally on the first move, what is the correct response?

6. (a) Come up with the winning strategy for Chomp with an *n*-by-2 board.
(b) Come up with the winning strategy for Chomp with an *n*-by-*n* board.

7. Use the work-backwards method to find all safe positions in 4-by-3 Chomp. What is the winning first move in this game? ✓

8. Is the Nim position with four piles of 3, 4, 5, and 6 stones safe? ✓

9. What is the correct move in Nim if three piles are left with 6, 8, and 11 stones in them? ✓

10. Write and test a Python function that takes a Nim position, represented as a list of non-negative integers, and checks whether it is safe.

11.♦ Consider the following modified version of Nim. In this game, stacks of coins are placed on some of the squares of a one-dimensional board:

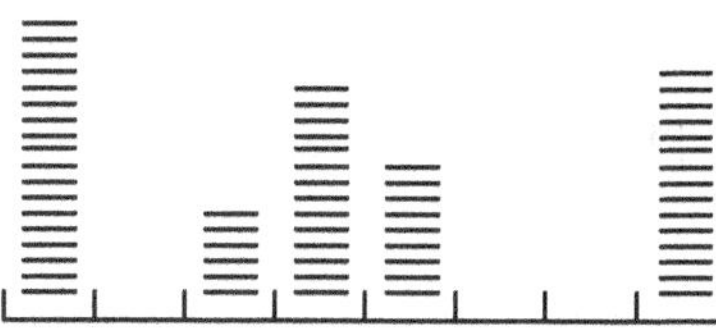

A player can take several coins (at least one) from any stack and add them to the next stack to the right (or start a new stack there, if that square was empty). Players are not allowed to take coins from the rightmost stack. Whoever moves the last coin wins. Describe the safe positions in this game.

12.♦ Six stacks of coins are arranged in a line on the table:

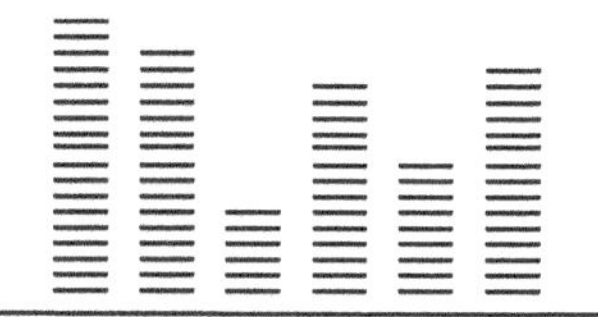

Two players take turns taking coins: a player must take a whole stack of coins, either on the left or on the right end. The player who ends up with most coins wins. Come up with a strategy for the first player that assures that he collects at least as many coins as his opponent. ⸱ Hint: imagine that the stacks are arranged on a chessboard, on squares of alternating colors. ⸱ ✓

13.■ In this game there are nine cards with the numbers 1 through 9 written on them. Two players take turns, taking one card on each turn. The player who is first to collect three cards with numbers that add up to 15 wins. If the first player takes 5, is 3 or 4 the correct response? Come up with a strategy that assures a win or a tie for the first player. ⸱ Hint: imagine that the cards are arranged on the table and form a 3-by-3 magic square (the sums of the numbers in each row, each column, and each of the two diagonals are the same); instead of picking up a card, the player writes her or his initials on it. ⸱

11.5 Review

Terms introduced in this chapter:

Redundancy
Parity
Checksum
Check digit
Substitution error
Transposition error
Invariant
Loop invariant
Combinatorial game
Safe and unsafe positions
Strategy stealing
Nim
Nim sum

'chapter' $\times$ '12'

c1 h1 a1 p1 t1 e1 r1

c2 h2 a2 p2 t2 e2 r2

Counting

12.1 Prologue

Combinatorics is a branch of mathematics that deals with counting all possible combinations, arrangements, sequences, or sets of objects. For example: Ice cream comes in two sizes, three flavors, and with any one of five toppings; in how many different ways can you order an ice cream? It is often not feasible simply to list and count all possible arrangements or combinations; combinatorics offers more sophisticated methods of counting. Many of these methods are not very complicated: they basically use the four arithmetic operations. The trick is to know when to multiply, when to divide, when to add, and when to subtract.

Counting methods came to prominence in the 17th century when Blaise Pascal (1623-1662) and others got interested in computing the odds in gambling games. For example, what are the odds of getting 11 when we roll two dice? To find out, we need to know the number of all favorable outcomes (possible rolls that give 11) and the number of all possible outcomes. Using combinatorial counting techniques, we can analyze the likelihood of different arrangements and outcomes in card games, such as poker and blackjack, dice games, such as craps, and so on. We can also analyze the complexity of computer algorithms and the running time and required space for computer programs.

12.2 The Multiplication Rule

The key operation in combinatorics is multiplication.

> **Suppose an object is described by two independent attributes. The total number of possible combinations of their values is equal to the number of possible values for the first attribute <u>times</u> the number of possible values for the second attribute.**

The multiplication rule is illustrated in Figure 12-1.

**Figure 12-1. Three mouth shapes and two colors
make six possible faces.**

If an object has three, four, or k attributes, then we multiply three, four, or k numbers, respectively.

Example 1

How many two-digit positive integers are there?

Solution

We can choose the tens digit in 9 ways (1 through 9) and the units digit in 10 ways. The answer is $9 \cdot 10 = 90$.

Example 2

A license plate has three digits followed by three letters. Suppose all combinations of digits and letters are allowed. What is the total number of possible license plate numbers?

Solution

There are 10 ways to choose the first digit, 10 for the second digit, and 10 for the third digit. There are 26 ways to choose the first letter; same for the second letter and the third letter. The answer is $10 \cdot 10 \cdot 10 \cdot 26 \cdot 26 \cdot 26 = 17,576,000$.

Example 3

How many different values can be represented in one byte?

Solution

There are 2 ways to set the first bit, 2 ways to set the second bit, and so on. The answer is $2^8 = 256$.

The multiplication rule explains how very large numbers can easily arise in combinatorial problems — or why it is so hard to win the lottery.

Example 4

How many ways are there to cover a whole 19 by 19 Go board with white and black stones?

Solution

2^{361} — more than 2×10^{120}, that is, many times more than there are atoms in the universe.

Section 12.2 ~ Exercises

1.   A combination lock has three wheels with the digits 0 through 9 on each wheel. What is the total number of possible combinations? ✓

2.   Ice cream comes in two sizes, three flavors, and with any one of five toppings. In how many different variations can you order an ice cream?

3.   How many three-letter names are possible in Python such that all letters are lowercase, the middle letter is a vowel ('a', 'e', 'i', 'o', 'u'), and the other two letters are consonants? ✓

4.   An experiment consists of tossing a coin 10 times. The outcome is recorded as a string of ten letters, either "H" or "T". How many different outcome strings are possible?

5.▪   In how many ways can we split 10 different marbles between two kids?
        ⸨ Hint: see Question 4. ⸩

6.   What is the number of all possible subsets of a set of ten elements, including the empty set and the whole set? 〈 Hint: see Question 5. 〉

7.   What is the number of all possible colors that can be shown on a computer screen if the graphics adapter generates a color using 8 bits for each of the red, green, and blue components of the color? ✓

8.   What is the number of all possible configurations of five disks of different sizes on three pegs in the Tower of Hanoi puzzle? Remember that we are not allowed to place a bigger disk on top of a smaller one. ✓

12.3 Permutations

The multiplication rule also applies when we want to count possible arrangements of objects without repetition.

Example 1

In how many ways can a coach choose three players — a point guard, a shooting guard, and a center — from a team of seven players?

Solution

If they came from three different teams, we could choose the first player in 7 ways, the second player in 7 ways, and the third player in 7 ways. But here we cannot just multiply $7 \cdot 7 \cdot 7$ because all three players come from the same team. Once we have chosen the first player, there are only 6 players left. So we can choose the second player in 6 different ways. Once we have done that, there are only 5 players left, so we can choose the third player in 5 different ways. The answer is $7 \cdot 6 \cdot 5$.

Example 2

How many ways are there to seat 6 students in a classroom with 20 desks?

Solution

We can seat the first student at any one of the 20 desks, the second student at any one of the remaining 19 desks, and so on. The answer is $20 \cdot 19 \cdot 18 \cdot 17 \cdot 16 \cdot 15 = 27,907,200$.

❖   ❖   ❖

> **An ordering of n different objects or symbols is called a *permutation*. There are $n \cdot (n-1) \cdot ... \cdot 2 \cdot 1 = n!$ possible permutations of n objects.**

We can choose the object for the first position in n ways, for the second position in $n-1$ ways, and so on. Only one object remains for the last position.

> **Recall that the product $1 \cdot 2 \cdot ... \cdot n = n \cdot (n-1) \cdot ... \cdot 1$ is called "n-factorial." It is denoted by $n!$.**

It is possible to arrive at the above result in a different way. Let P_n be the number of permutations of n objects. Let's take n objects, call one of them x and set it aside. There are P_{n-1} permutations of the remaining $n-1$ objects. For each of these P_{n-1} permutations, we can insert x at the beginning, between any two objects, or at the end, thus creating n different permutations of n objects. So from each of the P_{n-1} permutations of $n-1$ objects, we have now obtained n different permutations of n objects. Therefore, $P_n = nP_{n-1}$. Also, $P_1 = 1$, because there is only one permutation of 1 object. It is now clear that $P_n = n!$. To complete the proof more formally we would need to refer to *mathematical induction*, which is explained in Chapter 16.

Example 3

How many ways are there to put 5 different party hats on 5 people?

Solution

$5! = 120$.

Example 4

How many ways are there to order the cards in a deck of 52 cards?

Solution

$52! = 80658175170943878571660636856403766975289505440883277824000000000000$.

Section 12.3 ~ Exercises

1.   How many three-digit integers have all different digits? ✓

2.■ How many five-letter "words" (strings of letters) do not have the same letter twice in a row? Count any combination of letters as a "word."

3.   How many ways are there to choose "most likely to succeed," "best athlete," "best dancer," "best dressed," and "class clown," a boy and a girl in each category, in a graduating middle school class of 30 girls and 36 boys? The same person cannot be chosen in two categories.

4.   How many ways are there to seat 6 people in a row of 8 chairs? ✓

5.■ How many ways are there to seat 8 people in a row of 6 chairs, with two people left out? ⸨ Hint: assign chairs to people, not people to chairs. ⸩ ✓

6.   How many pages will it take to print all possible permutations of letters "ABCDEFGHIJKL"? Each permutation is printed on a separate line; 60 lines fit on a page.

7.■ Amelia is solving the cryptarithmetic puzzle

```
    S E N D
  + M O R E
  ---------
  M O N E Y
```

Her method is to try all possible substitutions of different digits for different letters (excluding 0 for 'M' and 'S', because a number cannot start with a 0). If she takes one minute for each try (her addition skills are excellent) and finds the answer after trying half of all possible substitutions, how long will it take Amelia to solve the puzzle? ✓

8.   Two words are called *anagrams* of each other if they are made up of the same letters used in a different order: for example, MANGO and AMONG. Jesse decided to write a program that finds all the anagrams of a given word by generating all possible permutations of its letters and looking up each permutation in a list of words (obtained in a file). If the program can generate and look up 360 permutations per second, how long will it take it to find all anagrams of "BINARY"? "DECIMAL"? What about "CONVERSATION"? ✓

12.4 Using Division

In many combinatorial problems it is easier first to overcount, counting each arrangement several times. This is fine, as long as we count each arrangement the same number of times and know that number. Then we can get the answer by dividing our total count by the number of times we counted each arrangement.

Example 1

n teams are playing in a round-robin tournament; each team plays once against every other team. How many games will be played?

Solution

There are n ways to choose the first team and $(n-1)$ ways to choose the second team for a game. That gives $n(n-1)$. However, this way we have counted each game twice, because Team A against Team B is the same game as Team B against Team A. Therefore, the answer is $\dfrac{n(n-1)}{2}$.

Example 2

How many seven-letter words can we make from the letters A L A B A M A (assuming any arrangement of letters is a "word")?

Solution

If all seven letters were different, we would have $7 \cdot 6 \cdot 5 \cdot 4 \cdot 3 \cdot 2 \cdot 1 = 7!$ words. But the four A's are the same, so we get the same word when we permute them. We have counted each arrangement of seven letters a number of times equal to the number of permutations of four A's, that is 4! times. The answer is $\dfrac{7!}{4!} = 5 \cdot 6 \cdot 7 = 210$.

Example 3

There are five identical pairs of gloves in a drawer. If you pull out two gloves at random, what are the odds that they will make a pair?

Solution

We can pull the first glove in 10 ways and the second glove in 9 ways. However, if the order is not important, we have counted each possibility twice. So the total number of ways to choose two gloves is $\dfrac{10 \cdot 9}{2} = 45$. We can pull a left glove in 5 ways and a right glove in 5 ways, so the number of ways to pull out a pair is $5 \cdot 5 = 25$. The number of ways to pull out 2 left gloves is $\dfrac{5 \cdot 4}{2} = 10$. The same for two right gloves. The number of ways to pull out two gloves that don't make a pair is $10 + 10 = 20$. Notice that the numbers match: $25 + 20 = 45$. The odds of getting a pair are $25 : 20 = 5 : 4$ (we say, "5 to 4").

❖　❖　❖

Sometimes, a problem can be solved either by division or by "anchoring" specific properties of an object or an arrangement.

Example 4

How many ways are there to seat a teacher and six students at a round table? Arrangements are considered the same as long as each person has the same left-hand and right-hand neighbors.

Solution

We can say that there are 7! total arrangements and 7 ways to rotate the table, so the answer is equal to $\dfrac{7!}{7} = 6! = 720$.

Another solution may be obtained by seating the teacher anywhere at the table; then there are 6! ways to arrange the students.

Section 12.4 ~ Exercises

1.   In how many ways can we choose two puppies out of a litter of 12 puppies? ✓

2.   Ice cream comes in two sizes, three flavors, and with any <u>two</u> of five toppings. How many different orders are possible?

3.■   How many words can we make by rearranging the letters T E N N E S S E E (assuming any arrangement of letters is a "word")?

4.■   How many ways are there to seat a group of six at a rectangular dinner table? The host and the hostess must sit at the short sides and the four guests two at each of the longer sides. Arrangements are considered the same as long as each guest has the same neighbors. ✓

5.■   In how many ways can we color the six faces of a cube in six different colors? The cube can be rotated any way you want — the colorings are considered the same. ✓

6.♦   An airline is planning three non-stop flights from the East Coast of the United States to the Caribbean. The airline serves six major cities on the East Coast and 12 different Caribbean islands. Multiple flights can go to the same island, but not from the same city. How many different configurations of the three flights are possible? ✓

7.♦   How many ways are there to place eight rooks on a chessboard so that none of them threatens any other? A rook moves vertically or horizontally by any number of squares. (The 64 squares on a chessboard are marked "a1" through "h8" and considered different; the eight rooks are considered identical.) ✓

12.5 Combinations

Combinations are selections of k elements from a given set of n elements ($0 \leq k \leq n$), disregarding the order. This number is written $\binom{n}{k}$ and read "*n*-choose-*k*."

$\binom{n}{k}$ are important numbers in mathematics, and they have many nice properties.

First of all, notice that there is a kind of symmetry: $\binom{n}{k} = \binom{n}{n-k}$. Indeed, to choose k objects is the same as to set aside the remaining $n-k$ objects. Note that $\binom{n}{n} = 1$.

Mathematicians have agreed that $\binom{n}{0} = 1$, so the symmetry is complete.

❖   ❖   ❖

Let's derive a formula (with factorials) for n-choose-k using the multiplication and division rules. Suppose we choose k objects as follows: we first arrange all n objects in line, then take the first k. The total number of arrangements is $n!$. However, if we rearrange the first k objects and/or the remaining $n-k$ objects, we will end up with the same selection. There are $k!$ ways to rearrange the first k objects and $(n-k)!$ ways to rearrange the remaining $n-k$ objects. So we have counted each selection $k!(n-k)!$ times. Therefore,

$$\binom{n}{k} = \frac{n!}{k!(n-k)!} = \frac{n \cdot (n-1) \cdot \ldots \cdot (n-k+1)}{k \cdot (k-1) \cdot \ldots \cdot 1}$$

It is convenient to define $0!$ as 1 so that the formula works for $k = 0$, too.

❖   ❖   ❖

It is useful to remember the formulas for $\binom{n}{k}$ for $k = 0$, 1, 2, and 3:

$$\binom{n}{0} = 1$$

$$\binom{n}{1} = n$$

$$\binom{n}{2} = \frac{n(n-1)}{2}$$

$$\binom{n}{3} = \frac{n(n-1)(n-2)}{6}$$

Example 1

How many ways are there for two children to share 4 different pencils, so that each gets two pencils?

Solution

We need to choose two pencils for the first child — the second child gets the other two.

$$\binom{4}{2} = \frac{4 \cdot 3}{2} = 6 .$$

Example 2

How many ways are there to choose 5 cards from a deck of 52 cards?

Solution

$$\binom{52}{5} = \frac{52 \cdot 51 \cdot 50 \cdot 49 \cdot 48}{5 \cdot 4 \cdot 3 \cdot 2 \cdot 1} = 2,598,960 .$$

Example 3

How many different bit patterns in a byte have exactly three bits set?

Solution

$$\binom{8}{3} = \frac{8 \cdot 7 \cdot 6}{3!} = 56 \,.$$

❖   ❖   ❖

$\binom{n}{k} = \dfrac{n!}{k!(n-k)!}$ is a neat formula, and it confirms the symmetry $\binom{n}{k} = \binom{n}{n-k}$. It is not practical for some computations, though, because factorials get very large quickly. For computer programs, it is more convenient to rewrite $\binom{n}{k}$ as a product of fractions:

$$\binom{n}{k} = \frac{n}{k} \cdot \frac{n-1}{k-1} \cdot \ldots \cdot \frac{n-k+1}{1}$$

Section 12.5 ~ Exercises

1. How many ways are there to order a pizza with three toppings out of five possible toppings? ✓

2. Write $\binom{4}{k}$ for $k = 0, 1, 2, 3, 4$.

3. Pat has 40 beads of which 35 are white and 5 are black. The beads are identical, except for the color. How many different bracelets can Pat make using all 40 beads? The ends of a bracelet are asymmetrical: one has a hook and the other an eyelet. ✓

4. ▪   Write a Python function `n_choose_k` that calculates and returns $\binom{n}{k}$ as an
int. The function should work for $0 \le k \le n$. ⸮ Hint: use a `float` for the
product of fractions, then convert it into an `int` using the built-in function
`round`. Rounding helps avoid tiny inaccuracies in computer arithmetic. ⸮

5. ▪   How many ways are there to split 9 different stickers among three people,
three apiece? ✓

6. ▪   How many different tic-tac-toe grids have 3 X's and 3 O's? ⸮ Hint: see
Question 5. ⸮

7. ◆   How many ways are there to make a "full house" poker hand from a deck of
52 cards? A full house is three cards of one rank and a pair of another
rank. ✓

8. ◆   If you multiply out $\underbrace{(x+1) \cdot \ldots \cdot (x+1)}_{n \text{ times}}$, what are the coefficients $c_0, c_1, \ldots, c_n$ of

the resulting polynomial $c_n x^n + c_{n-1} x^{n-1} + \ldots + c_1 x + c_0$?

9. ◆   Consider the following Boolean expression:

$$((P \text{ and } Q) \text{ or not } (P \text{ or } Q)) \text{ and } ((P \text{ and not } Q) \text{ or not } P)$$

It has three "and" and three "or" operators. Suppose we reshuffle these six
operators to make new expressions. Show that among all the different
expressions obtained this way, at least two will have the same truth tables.
⸮ Hint: don't do logic — just count. ⸮ ✓

12.6 Using Addition and Subtraction

Addition comes into play when it is easier to split the set of all possible objects or arrangements into two or more disjoint sets, count the arrangements in each of these sets separately, and then add the results.

Example 1

How many ways are there to choose one or two ice cream toppings from five available toppings?

Solution

One topping is not the same as two toppings! There are 5 ways to choose one and $\frac{5 \cdot 4}{2} = 10$ ways to choose two. The answer is $5 + 10 = 15$.

Example 2

How many ways are there to schedule two quizzes in a five-day week, if the quizzes cannot be on the same day or on consecutive days?

Solution

If the first quiz is on the first day, the second quiz can be on the third, fourth or fifth day — three possibilities. If the first quiz is on the second day, the second quiz can be on the fourth or fifth day — 2 possibilities. If the first quiz is on the third day, the second quiz must be on the fifth day — 1 possibility. The answer is $3 + 2 + 1 = 6$.

❖   ❖   ❖

If you have trouble counting objects or arrangements that satisfy a certain condition, you may try to count <u>all</u> possible arrangements, then count the arrangements that <u>do not</u> satisfy the condition and subtract their number from the total.

Example 3

How many ways are there to seat two adults and four kids in a row in a movie theater so that the two adults are not next to each other?

Solution

The total possible number of arrangements is 6! = 720. Let's count the number of arrangements where the adults <u>are</u> sitting together. The number of ways to choose two seats together for the adults is 5; the number of ways to seat the adults in these seats is 2; the number of ways to seat the kids in the remaining four seats is 4!. So the number of arrangements where the adults sit together is $5 \cdot 2 \cdot (4!) = 240$. The answer is $720 - 240 = 480$.

Section 12.6 ~ Exercises

1.   How many triangles are there in the following picture?

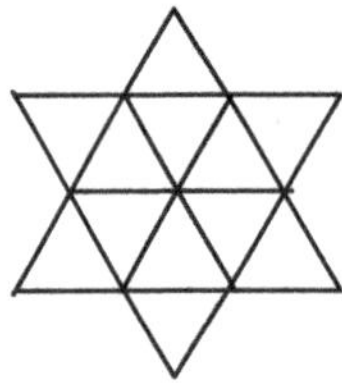

{ Hint: count triangles of different sizes separately. } ✓

2.   How many different ways are there to make 50 cents using quarters, dimes, and nickels?

3.   How many pairs of integers m and n, such that $2 \leq m < n \leq 12$, are relatively prime? Relatively prime numbers have no common factors except 1. ✓

4.   Recall that a valid name in Python can consist of upper- and lowercase letters, digits, and underscore characters, but cannot start with a digit. How many valid Python names of length 3 or less are there?

5. ▪ The Andover co-ed indoor soccer league tournament requires a 7-member team with at least two women. The Andover Pythons club has 8 men and 5 women. How many ways do the Pythons have to form a tournament team?

6. ♦ Given cards 2 through 10 in four suits (36 cards total), how many ways are there to choose three cards that add up to 21? ⸴ Hint: consider separately the cases where all three cards have the same rank, two cards have the same rank, and all three cards have different ranks. ⸲ ✓

7. How many four-digit numbers have at least one 7 among their digits? ⸴ Hint: count the numbers that do not have any 7s. ⸲ ✓

8. How many positive integers below 1000 are not divisible by 6?

9. ♦ In the diagram below, how many different paths lead from *A* to *B*?

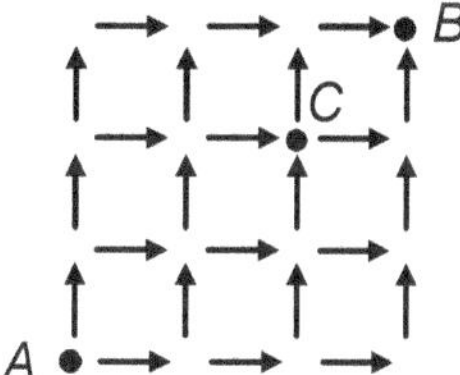

How many of them do not go through point *C*? ✓

10. ♦ A password can contain upper- and lowercase letters as well as digits; it must include at least one uppercase letter and at least one digit. How many different four- or five-character passwords are possible? ✓

12.7 Review

Terms and notation introduced in this chapter:

Combinatorics
Multiplication rule
Permutations
Factorial
Combinations
n-Choose-k

$$1 \cdot 2 \cdot \ldots \cdot n = n!$$

$$\binom{n}{k} = \frac{n!}{k!(n-k)!}$$

```
>>> chapter = choice([13, 14, 15])
>>> chapter
13
```

Probabilities

13.1 Prologue

When we roll a pair of dice, what is the likelihood of getting a sum of 11 points? Which is more likely, 7 or 11? If we roll two dice 100 times, approximately how many times will we get 11? Probability theory helps answer such questions.

The probability of an event is a number between 0 and 1 that describes the likelihood of that event happening when we repeat an experiment many times. Zero means the event never happens; 1 means the event always happens, and 0.5 means the event happens roughly half of the time.

To ask and answer questions about probabilities, we first need to describe formally what an event is; then we can figure out a way to calculate the probability of an event.

Suppose we have an experiment that produces some outcome (a number, an object, or a particular combination of objects). The set of all possible outcomes is called the *probability space* for the experiment. In this book, we will deal only with experiments that have a finite number of possible outcomes. In other words, the probability space is a finite set. For example, when we roll a die with 1, 2, 3, 4, 5, and 6 on its faces, there are six possible outcomes, so the probability space is a set of six elements, $\{1, 2, 3, 4, 5, 6\}$

An *event* is defined as a subset of the probability space that consists of all "favorable" outcomes that meet certain criteria. For example, we can define the following event: the number of points on the die is 3 or more. This event can be described as the set $\{3, 4, 5, 6\}$. Other examples of events: the number of points on the die is an even number: $\{2, 4, 6\}$; the number of points is 6: $\{6\}$. In the latter case, the event is a subset of the probability space that has only one element.

13.2 Calculating Probabilities by Counting

When all outcomes in the probability space are <u>equally likely</u>, the probability of an event is defined as the ratio of the number of favorable outcomes to the number of all possible outcomes:

$$\text{The probability of an event} = \frac{\text{The number of favorable outcomes}}{\text{The number of all possible outcomes}}$$

Example 1

When we roll a six-sided die with 1, 2, 3, 4, 5, and 6 on its faces, what is the probability of getting 3 or more?

Solution

There are 6 possible outcomes. All six are equally likely (assuming, of course, that the die is a perfect cube and not "loaded"). Four of the outcomes give 3 points or more: $\{3, 4, 5, 6\}$. Therefore, the probability of getting 3 or more is $\dfrac{4}{6} = \dfrac{2}{3}$.

The situation is a little more complicated when we roll a pair of dice. How do we define the probability space here? Let us say that any combination of points on the first die and on the second die is a different outcome. Then the probability space will consist of 36 elements (Figure 13-1). Again, each of the 36 different outcomes is equally likely.

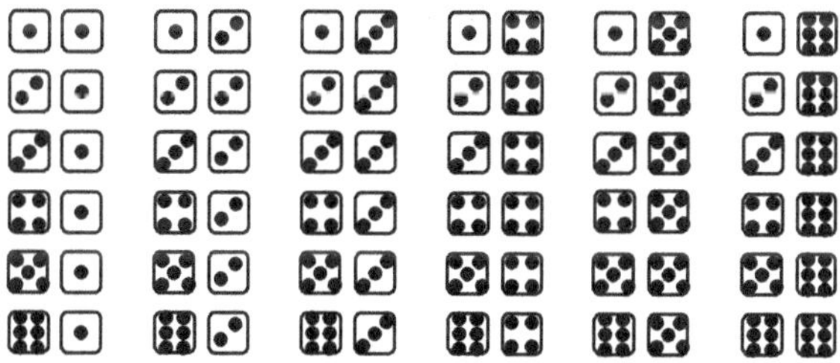

Figure 13-1. The probability space for possible rolls of two dice has 36 elements — all possible combinations of points on the first and the second die.

Example 2

When we roll a pair of dice, what is the probability that the sum of the points on the dice is 11? What is the probability of getting 7?

Solution

The probability of getting 11 can be calculated as 2/36 = 1/18, because exactly two possible outcomes produce a sum of 11 points: 5+6 and 6+5. The probability of getting a 7 is 6/36 = 1/6 because six possible outcomes give a sum of 7 points: 1+6, 2+5, 3+4, 4+3, 5+2, and 6+1 (Figure 13-2).

When the whole probability space is split into several non-overlapping events, the sum of the probabilities of these events is always 1.

Figure 13-2 shows an example.

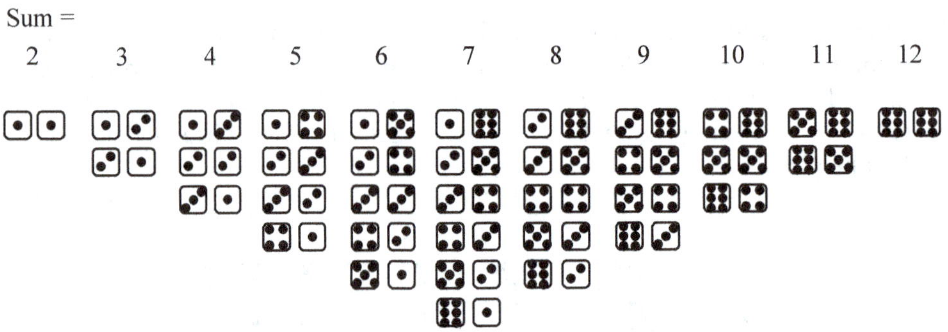

$$\frac{1}{36} + \frac{2}{36} + \frac{3}{36} + \frac{4}{36} + \frac{5}{36} + \frac{6}{36} + \frac{5}{36} + \frac{4}{36} + \frac{3}{36} + \frac{2}{36} + \frac{1}{36} = \frac{36}{36} = 1$$

Figure 13-2. Probabilities of getting a certain sum of points on a pair of dice.

You might be wondering: If we are interested in the sum of points on two dice, why not choose a simpler probability space that contains only 11 elements: the possible sums 2, 3, ..., 12? Technically we could, but such a probability space wouldn't be very useful, because, as we have seen, different outcomes in this space have <u>different</u> probabilities. This probability space wouldn't help us calculate the probability of each outcome, nor the probabilities of various other events.

 **When we construct a probability space for a problem, we always try to choose a space in which all outcomes have <u>the same probability</u>.**

In such a space, the probability of each outcome is $1/n$, where n is the number of all possible outcomes. We then can find the probability of an event simply by counting the number of outcomes that meet the criteria for the event and dividing that count by n.

❖   ❖   ❖

Sometimes we can define the probability space in different ways for the same situation, but we should always get the same result for the probability of the same event.

Example 3

If we pull two random cards from a deck of 52 cards, what is the probability of getting two aces?

Solution

We can define the probability space as the set of all pairs of cards (disregarding the order of the two cards in the pair). There are $\binom{52}{2} = \dfrac{52 \cdot 51}{2}$ ways to choose a pair of cards from a deck of 52 cards. There are $\binom{4}{2} = \dfrac{4 \cdot 3}{2}$ ways to choose a pair of aces out of 4 aces (favorable outcomes). The answer is $\dfrac{4 \cdot 3}{2} \bigg/ \dfrac{52 \cdot 51}{2} = \dfrac{4 \cdot 3}{52 \cdot 51} = \dfrac{1}{221}$.

Alternative solution: we can define the probability space as the set of all <u>ordered</u> pairs of cards. The number of elements in this space is $52 \cdot 51$. The number of ordered pairs of aces is $4 \cdot 3$. We get the same answer: $\dfrac{4 \cdot 3}{52 \cdot 51} = \dfrac{1}{221}$.

Example 4

If we toss a coin three times, what is the probability of getting heads exactly twice?

Solution

The probability space here is {HHH, HHT, HTH, HTT, THH, THT, TTH, TTT}, and the desirable outcomes are {HHT, HTH, THH}. The answer is 3/8.

Section 13.2 ~ Exercises

1.   If you make a random guess, what is the probability of your guessing the right answer to a multiple-choice question with 5 answer options? ✓

2.   If you roll two dice, what is the probability that the sum of the points is greater than 7?

3.   What is the probability of winning the grand prize in a lottery where you have to mark the correct six squares on a grid of 36 squares? ✓

4.   If you toss a coin 5 times, what is the probability of getting tails at least 4 times?

5.   A roulette wheel has 36 slots numbered 1 through 36 plus 2 slots, marked zero, that win for the "house." The 36 slots alternate 18 red and 18 black; the zeros are green. What is the probability of winning if you bet on 17? What is the probability of winning if you bet "on red" (you win if the ball hits any red slot)? ✓

6.   What is the probability that a randomly selected number chosen from the first 50 positive integers is divisible by 3?

7.   What is the probability that the two digits in a randomly selected two-digit positive integer (that is, an integer between 10 and 99) are the same?

8.▪ Write a Python program that calculates the probabilities of getting different sums of points (from 3 to 18) when you roll <u>three</u> dice. What sum has the highest probability? ⋛ Hint: initialize a list of counts, all set to 0, then generate all possible combinations of points on three dice and for each combination increment the appropriate count. ⋛

13.3 More Probabilities by Counting

As we know, to calculate the probability of an event we have to count the number of all "favorable" outcomes and divide it by the number of all possible outcomes. When we do the counting, we can rely on all the tricks of the trade that we learned when practicing with the combinatorics problems in Chapter 12.

Example 1

If a random poker hand of five cards is dealt from a deck of 52 cards, what is the probability of getting "two pairs" (that is, a pair of the same rank, a pair of another rank, and any fifth card of a third rank)?

Solution

The total number of ways to deal a five-card hand is $\binom{52}{5} = \dfrac{52 \cdot 51 \cdot 50 \cdot 49 \cdot 48}{5 \cdot 4 \cdot 3 \cdot 2 \cdot 1} = 52 \cdot 51 \cdot 10 \cdot 49 \cdot 2$. To form "two pairs," let's first choose the 2 ranks of the pairs out of 13. There are $\dfrac{13 \cdot 12}{2} = 78$ ways to do that. Within each rank, there are 6 ways to choose 2 cards out of 4. Finally there are 44 ways to choose the remaining card (out of the remaining 11 ranks). The answer is $\dfrac{78 \cdot 6 \cdot 6 \cdot 44}{52 \cdot 51 \cdot 10 \cdot 49 \cdot 2} = \dfrac{198}{4165} \approx 0.048$.

Example 2

There are 60 jelly beans in a bag: 15 red, 15 blue, 15 green, and 15 yellow. Danny has pulled out one red and one blue, and 58 jelly beans remain in the bag. If Danny pulls out three more jelly beans at random, what is the probability that he will end up with exactly three jelly beans of the same color?

Solution:

We will use the addition and multiplication rules of counting. There are four mutually exclusive possibilities for a favorable outcome: Danny can end up with three red, three blue, three green, or three yellow jelly beans. To get three red, he needs to choose 2 of the 14 red jelly beans remaining in the bag. There are $\binom{14}{2} = \frac{14 \cdot 13}{2} = 91$ ways of doing that. Then he needs to choose one more jelly bean, not red. There are $58 - 14 = 44$ ways of doing that. Therefore, there are $91 \cdot 44 = 4004$ ways of getting five jelly beans with three reds among them. The same for three blue jelly beans. To get three green jelly beans, Danny needs to choose 3 out of 15. There are $\binom{15}{3} = \frac{15 \cdot 14 \cdot 13}{3 \cdot 2} = 455$ ways of doing that. Similarly, there are 455 ways to choose three yellow jelly beans. The total number of favorable outcomes is $4004 + 4004 + 455 + 455 = 8918$. The total number of all possible outcomes is the number of ways to choose 3 jelly beans out of 58, which is $\binom{58}{3} = \frac{58 \cdot 57 \cdot 56}{3 \cdot 2 \cdot 1} = 30856$. The probability of getting exactly three of the same color is $\frac{8918}{30856} \approx 0.289$.

When we calculate the probability as a ratio of the number of favorable outcomes to the number of all possible outcomes, it is very important that all possible outcomes in the probability space have equal probabilities. If that is not the case, then the simple ratio method will result in a wrong answer.

Example 3

Consider a game between two players. Each tosses a coin up to three times. Whoever gets heads first wins; if they get the same side of the coin, they play a tiebreaker; if they get the same side again, they play another tiebreaker. What is the probability that the first player will win?

Solution

In a naive approach we would consider all possible runs of the game:

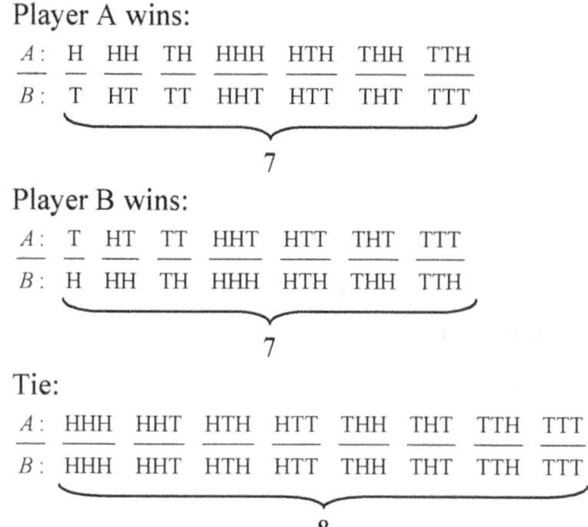

There are seven possible ways in which Player A could win, and similarly seven possible ways in which Player B could win. There are eight possible ways of the game ending in a tie.

It seems that the probability of Player A winning is $\dfrac{7}{7+7+8} = \dfrac{7}{22}$. Wrong answer!

The 22 possible runs of the game, listed above, are <u>not</u> equally likely. To see this, let us allow each player to toss three times, no matter what, and only then determine the winner:

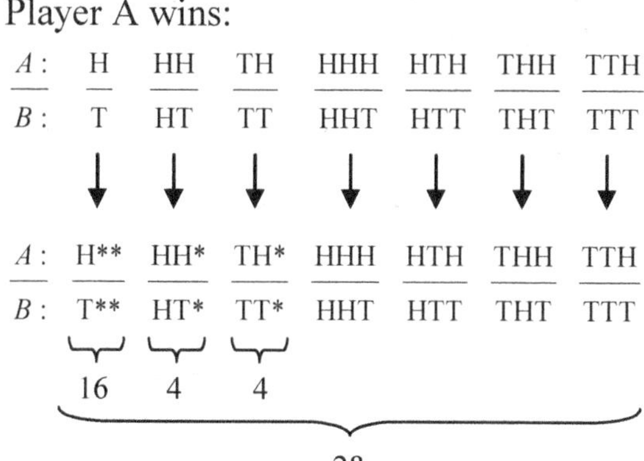

Here * is a wildcard character that can stand for either side of the coin. This new probability space is properly defined, with all outcomes equally likely. The old space did not have this property: as we can see now, the H/T game was actually four times more likely than the HH/HT game and 16 times more likely than the HHH/HHT game.

The total number of possible outcomes in the new space is $2^6 = 64$; the correct answer is $\dfrac{28}{64} = \dfrac{7}{16}$.

In this example we did not have to count all the winning combinations: there is a shortcut based on the symmetry of the players. The total number of outcomes is 2^6; the number of tie-game outcomes is 2^3; so the number of outcomes where Player A wins is $\dfrac{\left(2^6 - 2^3\right)}{2} = 28$.

Section 13.3 ~ Exercises

1.   What is the probability of getting a "four of a kind" poker hand (four cards of the same rank plus any fifth card)? ✓

2.■   A "royal flush" is the best possible poker hand: ace, king, queen, jack, and ten in the same suit. What is the probability of getting a royal flush when you are randomly dealt five cards from a standard deck of 52 cards? ✓

3.■   In the game of "ten-spot" Keno you mark 10 "spots" (numbers) on a playing card that has 80 numbers. The "house" (that is, the casino) then picks 20 numbers randomly. To win the "grand prize" you need to have all 10 of your numbers "hit" by the house numbers. What is the probability of winning the grand prize? (The game of Keno was brought to the United States from China and quickly became popular in casinos, gambling establishments, and even state lotteries. This game gives the player the <u>worst</u> odds of winning of any casino game.) ✓

4.■   A bag has 3 red marbles and 5 blue marbles. If you randomly pull out 4 marbles, what is the probability of getting two red ones and two blue ones?

5.◆   You draw cards from a deck of 52 cards until you get two cards of the same rank. What is the probability that you'll end up with no more than three cards? ✓

6.◆   The pages in a book are numbered from 1 to 96. We cut out all the page numbers and cut them into individual digits. What is the probability that the sum of two digits chosen at random from this pile will be 10?

7.◆   Sixteen billiard balls are randomly dropped into 6 pockets. (Pretend a pocket can hold any number of balls.) What is the probability that the northeast corner pocket has fewer than three balls? ✓

8.◆   What is the probability that a random Blackjack hand of 3 cards is worth 21 points? In Blackjack, each number card 2 through 10 counts at face value; any ace counts as 1 or 11, your choice; any other "picture" card (jack, queen, or king) counts as 10. ✓

13.4 Multiplication, Addition, and Subtraction

Remember how we used multiplication, addition, and subtraction for counting? We can, of course, apply these methods when we count "favorable" outcomes and all possible outcomes of an experiment. But there is a shortcut: we can apply multiplication, addition, and subtraction directly to probabilities. Let's start with multiplication.

> **Suppose we have two experiments that are independent of each other. Suppose event *A* can occur in the first experiment and event *B* can occur in the second experiment. The probability that both *A* and *B* happen in their respective experiments is the probability of *A* times the probability of *B*.**

Example 1

What is the probability of getting two sixes when we roll a die twice in a row?

Solution

The probability of getting a six on the first roll is $\dfrac{1}{6}$ and the probability of getting a six on the second roll is $\dfrac{1}{6}$. The probability of getting two sixes in a row is $\dfrac{1}{6} \cdot \dfrac{1}{6} = \dfrac{1}{36}$. Note that rolling a die twice is the same as rolling two identical dice together.

Example 2

I have three envelopes. Two of them hold one dollar bill; the third is empty. You are allowed to take any two envelopes. What is the probability that you will end up with two dollars?

Solution

When you take the first envelope, the probability of getting a dollar is $\dfrac{2}{3}$. If you got the first dollar, the probability of getting the second dollar from the remaining two envelopes is $\dfrac{1}{2}$. The probability of getting two dollars is $\dfrac{2}{3} \cdot \dfrac{1}{2} = \dfrac{1}{3}$. Here we assume that the first experiment was successful before we proceed with the second, but the multiplication rule still works. This is analogous to counting combinations without repetitions, as we discussed in Section 12.3.

❖   ❖   ❖

> **Probabilities can be multiplied in two circumstances: (1) When the events are independent from each other, as in Example 1 above, or (2) when the probability of the second event is calculated based on the assumption that the first event has already occurred, as in Example 2 above.**

Example 3

On the Caribbean island of Nevis it rains, on average, 45 days per year. What is the probability of rain on two consecutive days?

Solution

This is a trick question. You might be tempted to apply the multiplication rule and say that the probability is $\dfrac{45}{365} \cdot \dfrac{45}{365}$ or $\dfrac{45}{365} \cdot \dfrac{44}{364}$. However, the multiplication rule requires that the events happen <u>independently from each other</u>. Not so with weather: on Nevis the rainy months are September and October, and in those months rain on two consecutive days is quite likely. Suppose, for the sake of argument, that on some Mystery Island the 45 rainy days are always September 1 through October 15, and it never rains at all during the rest of the year. Then, if we choose two consecutive days in a year randomly, the probability of having rain on both will be $\dfrac{44}{365}$. We cannot answer the original question about Nevis without additional information about the distribution of rainy days throughout the year.

We use addition when we need to find the probability of an event that can be split into two non-overlapping events.

Example 4

In the casino game of craps, you roll two dice. If you roll 7 or 11, you win. If you roll 2, 3, or 12, you lose. Otherwise the game continues. What are the probabilities of winning and losing on the first roll?

Solution

The probability of winning on the first roll is $\dfrac{6}{36} + \dfrac{2}{36} = \dfrac{2}{9}$ (see Figure 13-2); the probability of losing on the first roll is $\dfrac{1}{36} + \dfrac{2}{36} + \dfrac{1}{36} = \dfrac{1}{9}$. (This is not the end of the game, though. The rules are such that ultimately the casino wins more than half of the games.)

Example 5

In a *random walk*, you choose the direction of each next step — forward, back, left, or right — randomly, with equal probabilities. What is the probability of returning to the starting position after two steps?

Solution

The following four sequences return to the starting position after two steps: {forward, back}, {back, forward}, {left, right}, and {right, left}. The probability of each sequence is $\dfrac{1}{4} \cdot \dfrac{1}{4} = \dfrac{1}{16}$. The probability that any one of the four sequences takes place is $\dfrac{1}{16} + \dfrac{1}{16} + \dfrac{1}{16} + \dfrac{1}{16} = \dfrac{1}{4}$.

> **If the probability of an event happening is p, the probability of the same event not happening is $1 - p$.**

Example 6

The probability of Dave hitting a home run is 0.13. What is the probability that Dave won't hit a home run?

Solution

$1 - 0.13 = 0.87$.

Example 7

Emily hits the "bull's eye" with a dart on average one out of fifteen times. What is the probability that Emily will hit the bull's eye at least once in three attempts?

Solution

Suppose the probability that Emily hits the bull's eye is p. Then the probability that she misses is $1-p$. The probability that she will miss three times in a row is $(1-p)^3$. The probability that she will hit the bull's eye at least once out of three attempts is $1-(1-p)^3$. Here $p = \dfrac{1}{15}$. The answer is $1-\left(1-\dfrac{1}{15}\right)^3 = \dfrac{631}{3375} \approx 0.19$.

Section 13.4 ~ Exercises

1. What is the probability of rolling 3 on a die three times in a row? ✓

2. In Scrabble™ there are 98 tiles with letters written on them. There are 2 Bs, 9 As, and 6 Ts. If you pull three tiles randomly out of the box, each time returning the tile to the box, what is the probability of getting B, A, T, in that order?

3. What is the probability of rolling a die 3 times and never getting a 6? What is the probability of getting a 6 at least once? ✓

4.■ What is the probability that in a random walk, as described in Example 5, you will return to the starting position after four steps?

5.■ In the game of squash, an unbroken sequence of successive strokes is called a rally. To win a point when you are serving, you need to win the rally. If your opponent served, winning a rally just gives you the right to serve next. If Ellie wins, on average, 4 out of 10 rallies when she serves and 3 out of 10 rallies when her opponent serves, what is the probability that Ellie will get a point after two rallies starting with the opponent's serve? ✓

6.♦ Question 5 describes the rules for scoring in squash. Susan beats Jimmy on average in 2 out of 3 rallies, regardless of who serves. If Susan starts out serving, what is the probability that she will win the next point within three rallies? Within five rallies? What is the probability that Susan will win the next point eventually? ⟨ Hint: in the latter case we get an infinite series, but we already know how to handle it... ⟩ ✓

7.◆   What is the probability that a player will win the next point in squash (see Question 6) if it is her serve and she is playing an equal opponent (that is, each player's probability of winning a rally is 0.5, regardless of who serves)?

8.◆   A dartboard has 20 outer sectors, 20 inner sectors, and a center called the "bull's eye." If, on average, out of 30 tries I hit the bull's eye once, an inner sector 5 times (equally likely among them), an outer sector 20 times, and miss the board completely 4 times, what is the probability that I will hit the same sector (or the bull's eye) twice in a row? ✓

9.◆   In the Tri-State Megabucks lottery you choose six numbers out of 42 numbers. If your numbers match the six drawn by the computer, you hit the jackpot. The computer also draws a seventh "bonus" number. If you didn't hit the jackpot but your six numbers match any six numbers out of seven drawn by the computer (including the bonus number), you get a consolation prize of $10,000. There are other prizes, but they are relatively small. You decide to play only if your average win (probability of jackpot times the size of jackpot plus probability of bonus times the size of bonus) exceeds the ticket cost. If a ticket costs $1.00, what jackpot size makes it worth playing for you?

10.◆   A pawn starts on a black square at the lower left corner of the chessboard. On each move the pawn moves one square up or to the right, with equal probabilities. It stops when it reaches the right or the top edge. What is the probability that the pawn will finish on a black square? ⋛ Hint: extend the board so that it forms an equilateral right triangle with the right angle at the lower left corner, and extend all paths to its hypotenuse, so that all paths become equal in length. ⋛

13.5   Pseudorandom Numbers

Computers are supposed to act predictably: start at the same position, perform the same steps, and you should get the same result. But sometimes we want the computer to act randomly. This is useful, for example, in games and in computer simulations of random processes.

A typical programming language has library functions that generate "random" numbers. The numbers are generated in software using a certain algorithm, so they are not really random, but they approximate random behavior. Such numbers are called *pseudorandom numbers*.

The Python library has a module `random`, which has many functions that generate and return random numbers. To obtain a random integer r in the range $a \le r \le b$, import the function `randint` from the module `random` and call `randint(a, b)`. For example:

```
>>> from random import randint
>>> randint(1,3)
2
>>> randint(1,3)
3
>>> randint(1,3)
3
>>> randint(1,3)
1
>>> randint(1,3)
2
```

(Your display may differ, since `randint` returns a pseudorandom number.)

The function `random` from the same module returns a `float` x in the range $0.0 \le x < 1.0$. For example:

```
>>> from random import random
>>> random()
0.215037630019111777
>>> random()
0.80958843480468412
```

Another function, `choice`, returns a randomly chosen character from a string (or element from a tuple, list, or any other "sequence"). For example:

```
>>> from random import choice
>>> choice('ABC')
'B'
>>> choice('ABC')
'A'
>>> choice('ABC')
'A'
```

The function `shuffle` rearranges the elements of a list in random order. For example:

```
>>> from random import shuffle
>>> lst = [1, 2, 3, 4, 5]
>>> shuffle(lst)
>>> lst
[5, 2, 3, 1, 4]
```

❖   ❖   ❖

Sometimes a theoretical solution gives us a formula, but the numbers may be too large or unwieldy to compute. Then a computer can help, either with calculations or by modeling the random process and observing the result. Such models are called *Monte Carlo simulations* (named after the town and popular gambling resort in Monaco).

Example 1

In a group of 25 people, which is more likely: that there are two people who share a birthday, or that everyone's birthday is on a different day?

Solution

Let's assume that a birthday can fall on any of the 365 days. (Ignore leap years.) One approach is to program a Monte Carlo simulation. Generate a random set of 25 birthdays and check whether any two fall on the same day. Repeat, say, 10,000 times, and count how many sets have shared birthdays and how many don't. This program is left to you as an exercise (Question 6).

Another approach to solving the Birthdays problem is to calculate the probability theoretically, using the methods that we have learned. The first person can have a birthday on any of the 365 days, the second person, too, and so on. Using the multiplication rule, we find that there are 365^{25} possible arrangements of birthdays. Using the multiplication without repetitions rule, we conclude that there are $365 \cdot 364 \cdot ... \cdot 341$ arrangements when all 25 people have different birthdays. Therefore, the probability that all people have different birthdays is $p = \dfrac{365 \cdot 364 \cdot ... \cdot 341}{365^{25}}$, and the probability that at least two people have the same birthday is $1 - p$.

Now let's answer the same question for n people. There is no extra work involved and we will be able to obtain the "crossing point": the number of people when the probability gets greater than 1/2. Try guessing that number now just to check your intuition!

It seems we have to deal with very large numbers for the numerator and the denominator. For example, 365^{25} has 65 digits. (We found that number using Python.) Python supports big integers, where a number can have any length, limited only by the size of the computer memory. When regular four-byte `int` values get out of range, Python automatically switches to using long values. So in Python, we can just calculate the top and the bottom of the fraction, take their ratio and be done. In other languages, however, handling very large numbers may be more involved. Luckily, we can handle this problem more economically, without big integers. Note that we can split the ratio into the product of individual fractions: $\dfrac{365 \cdot 364 \cdot \ldots \cdot 341}{365^{25}} = \dfrac{365}{365} \cdot \dfrac{364}{365} \cdot \ldots \cdot \dfrac{341}{365}$. It is easy to compute each fraction here as a `float`. In Question 5, we ask you to complete this program.

Section 13.5 ~ Exercises

1. Write and test a Python function that generates a hand of five cards chosen randomly from a deck of 52 cards. Each card is described by a pair (tuple) that holds the card's suit ('S', 'H', 'D', or 'C', for Spade, Heart, Diamond, or Club) and rank (1 through 13). ✓

2. Suppose you have a Python function `random_letter` that returns a random letter of the alphabet (with approximately equal probabilities for all 26 letters). What is the probability that `random_letter` will return three vowels (any of the letters 'A', 'E', 'I', 'O', 'U') in a row? ✓

3. Write a Python function `random_letter` as described in Question 2. Run it three times to see if you get three vowels in a row. Repeat 100,000 times, count how many times you got three vowels in a row, and estimate the probability of such an event. Compare your result to the theoretical probability that you obtained in Question 2.

4. Write a short program that uses a Monte Carlo method to estimate π. "Throw" a million random (virtual) points into a unit square and count for how many of them the distance from the origin is less than 1.

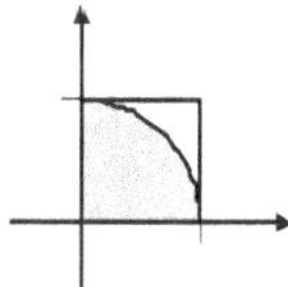

5. Using the theoretical approach described in this section, write a program that prints out the probabilities that at least two people in a group of n have the same birthday, for $n = 1$ through 50, without using big integers. This program is just a few lines of code. The output should look like this:

```
1:  0.000
2:  0.003
3:  0.008

. . .

48:  0.961
49:  0.966
50:  0.970
```

Watch out for "OBOBs" (off-by-one bugs), as we call the bugs that happen when a loop in your program runs one too few or one too many times. Determine from the printout the value of n at which the probability becomes greater than 0.5.

6. Write a program that implements a Monte Carlo simulation for birthdays. Write a function that generates a list of n random birthdays (numbers from 1 to 365) and returns `true` if all the numbers in the list are different. Call this function a sufficient number of times to estimate the probability that at least two people out of n have the same birthday. Compare your result with the theoretical result obtained in Question 5. ⸖ Hint: An easy way to find out whether generated numbers have duplicates is to allocate 365 counts and increment the appropriate count for each random number, then check whether any count is greater than 1. ⸖

7. In turtle graphics, draw a circle of radius 100 centered at the origin and return the turtle home (to the origin). Then simulate a random walk: on each move, set the turtle's direction to 0, 90, 180, or 270 degrees, chosen randomly with equal probabilities, then move forward by 20. Keep the pen down to see the turtle's path. Count and report the number of steps it takes the turtle to exit the circle. ⸖ Hint: the `turtle`'s function `distance(x,y)` returns the distance from the current turtle's position to the point (x, y). ⸖

8.■ The function below returns a random positive number from the "sequence" s:

```python
def positive_choice(s):
    lst = [x for x in s if x > 0]
    if len(lst) > 0:
        return choice(lst)
```

Each positive number in the list is equally likely to be chosen. Rewrite this function without creating any temporary list or set of positive numbers from s and without counting their number in advance. ⸙ Hint: Suppose r holds a positive number randomly chosen from the first k positive numbers in s. Start with k = 0 and r = None. When you get the next positive number x, you can either keep r unchanged or replace r with x. What should be the probability of replacing r with x? You never know: x may be the last positive number in s. You need to give it a fair chance... ⸙

Test your version of positive_choice by running it, say, 10000 times on (–1, 1, 2, 0, 3, –2, 4, 5, –6) and counting how many times each of the numbers 1, 2, 3, 4, 5 is returned. All the counts should be around 2000; it is highly unlikely that a count differs from 2000 by more than 120. ✓

9.■ The purpose of this exercise is to explore what happens when you add many random numbers together. Write a short program that helps you do that. Create a list of 100 counters: counters = 100*[0]. Run the following experiment: take the sum of 100 random numbers $0 \le r < 1$ (as returned by the random function), truncate the sum to an integer, getting i, and increment counters[i]. Repeat these steps, say, 2000 times. Then display the values in counters as vertical segments (a kind of simple bar chart) in turtle graphics. What does the resulting graph look like?

⸙ Hints:

```python
for i in range(100):
    x = x_offset + x_scale*i
    y = y_scale*counters[i]
    ...
```

Use reasonable offsets and scales, for example,

```python
x_offset, y_offset = -200, -100
x_scale, y_scale = 5, 1
```

Use goto(x, y) to draw.

⸙

13.6 Review

Terms introduced in this chapter:

> *Probability*
> *Probability space*
> *Independent events*
> *Multiplication rule for*
> *   probabilities*
> *Addition and subtraction*
> *   rules for probabilities*
> *Monte Carlo simulations*
> *Random walk*

Some of the Python features introduced in this chapter:

```
from random import randint, random, choice, shuffle
randint(m, n)
random()
choice(s)
shuffle(lst)
```

$$\begin{pmatrix} c & h & a \\ p & t & e \\ r & 1 & 4 \end{pmatrix}$$

Vectors and Matrices

14.1 Prologue

A *vector* is an entity that has a magnitude and direction. Vectors are widely used in physics. For example, if you want to describe the motion of an object at any given point in time, you might be interested in its speed (how fast it is going) and the direction of the motion. The velocity vector is a convenient way to combine these two characteristics (Figure 14-1-a). Similarly, when a force is applied to an object, it is convenient to represent it as a vector, because a force is characterized both by its magnitude and its direction. Figure 14-1-b shows the three forces that act on a block resting on a wedge: the gravitational force, the normal force, and the friction force.

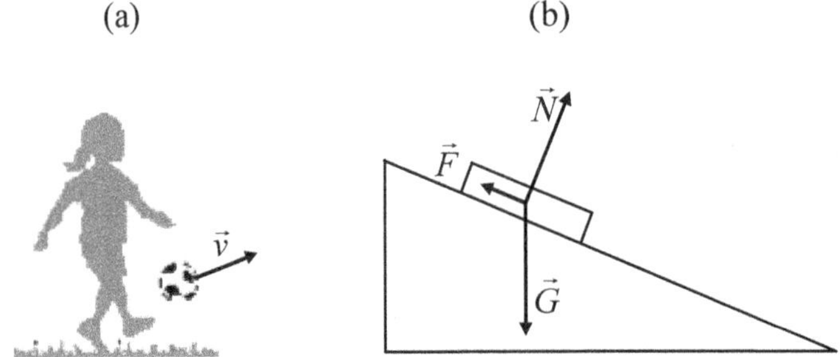

**Figure 14-1. (a) the velocity of an object is a vector
(b) force is a vector**

Another approach to vectors is more abstract. A two-dimensional vector is simply an ordered pair of real numbers, (x, y). A three-dimensional vector is an ordered triplet of numbers, (x, y, z). An n-dimensional vector is an *n-tuple* $(x_1, x_2, ..., x_n)$. Representing vectors using numbers is akin to representing points by their coordinates. This is convenient in formal computations and mathematical models.

There is a one-to-one correspondence between these two views of vectors. For example, in plane geometry, a vector can be viewed as an arrow from the origin to the point (x, y) on the plane (see Figure 14-2) or simply as a pair (x, y).

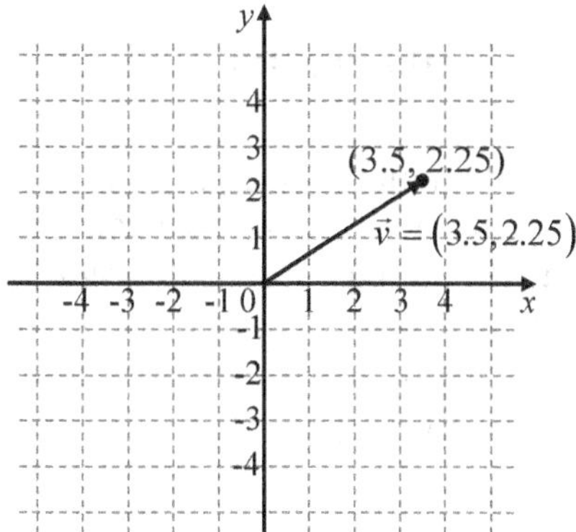

Figure 14-2. Correspondence between two-dimensional vectors and coordinates of points on the plane

In mathematics, a rectangular table of numbers is called a *matrix*. What can we do with vectors and matrices? At first it might seem like not much. A table of numbers is a matrix (Figure 14-3). Tables are useful for organizing and presenting data.

	2016	2017	2018	2019
1st Qtr	3,512	3,002	3,623	3,216
2nd Qtr	4,720	4,500	4,295	4,080
3rd Qtr	3,827	3,391	3,994	3,511
4th Qtr	5,008	4,856	4,659	4,388
Total	17,067	15,749	16,571	15,195

Figure 14-3. An example of a table

But this is not all, of course. A whole very rich branch of mathematics, called *linear algebra*, deals with vectors and matrices. Linear algebra plays an essential role not only in math but also in physics, engineering, and many other fields of science, and in computer graphics and 3D modeling. When you see a rotating 3D image in a design program or video game or special effects in a movie, you can be sure that some pretty sophisticated linear algebra algorithms are involved. Even such a trivial operation as changing the direction of a turtle in turtle graphics relies on basic linear algebra. In this chapter we will review some of the basic operations on vectors and matrices.

14.2 Operations on Vectors

Cartesian coordinates represent a point on the plane as a pair of numbers. They also allow us to represent a two-dimensional vector as a pair of numbers. The easiest way to see that is simply to draw any vector from the origin (Figure 14-2). So we can represent any two-dimensional vector as a pair of numbers (x, y), which are the coordinates of the end of the vector when its beginning is at the origin. The zero vector $\vec{0} = (0, 0)$ corresponds to the origin. Here we will focus on two-dimensional vectors and their representation as pairs of numbers.

> x **and** y **are called the** *components* **of the vector** (x, y).

> $|\vec{v}|$ **stands for the magnitude (length) of the vector** $\vec{v}$. **In two**
> **dimensions, the length is obtained by the Pythagorean theorem: for**
> $\vec{v} = (x, y)$, $|\vec{v}|^2 = x^2 + y^2$ $\Rightarrow$ $|\vec{v}| = \sqrt{x^2 + y^2}$.

For an n-dimensional vector, $\vec{v} = (x_1, x_2, ..., x_n)$, its length is defined as $|\vec{v}| = \sqrt{x_1^2 + x_2^2 + ... + x_n^2}$.

> **A vector of magnitude (length) 1 is called a** *unit vector.*

For any non-zero vector $\vec{u}$, $\dfrac{\vec{u}}{|\vec{u}|}$ is a unit vector.

❖   ❖   ❖

The beauty of the component representation of vectors is that vector arithmetic is straightforward. Given $\vec{u} = (x_1, y_1)$ and $\vec{v} = (x_2, y_2)$, and a number k,

> $\vec{u} + \vec{v} = (x_1 + x_2, y_1 + y_2)$
>
> $\vec{u} - \vec{v} = (x_1 - x_2, y_1 - y_2)$
>
> $k\vec{u} = (kx_1, ky_1)$

In other words, vectors can be added, subtracted, or multiplied by a number by performing the operation on the vector's components. For example:

$(7, 3) + (1, 10) = (8, 13)$

$(4.5, -2) - (1, 6.2) = (3.5, -8.2)$

$2 \cdot (2, 6) = (6, 12)$

Component-wise addition and subtraction and multiplication by a number on n-dimensional vectors obey the commutative and distributive laws:

$$\vec{u} + \vec{v} = \vec{v} + \vec{u}$$

$$\vec{u} + \vec{0} = \vec{u}$$

$$\vec{u} + (-\vec{u}) = \vec{0}$$

$$k(\vec{u} + \vec{v}) = k\vec{u} + k\vec{v}$$

Any set with such operations defined on its elements is called a *vector space* (or a *linear space*) over numbers. Naturally, a set of all n-dimensional vectors is a vector space over numbers (real numbers if the components are real numbers, or integers if the components are restricted to integers). A set of all polynomials of degrees that do not exceed n is another example of a vector space. But the set of literal strings with the operations of addition and multiplication by a number as defined in Python is not a vector space, because these operations obey neither the commutative nor the distributive law and because subtraction is not defined for strings.

> **In Python, an *n*-dimensional vector can be represented by a list or a tuple of *n* numbers.**

Example 1

```python
def sum_vectors(_u, _v):
    """Return the sum of two n-dimensional vectors _u and _v."""
    return [_u[i] + _v[i] for i in range(len(_u))]

def k_times_vector(_u, k):
    """Return the n-dimensional vector _u multiplied by k."""
    return [k*u for u in _u]
```

❖   ❖   ❖

Another essential operation on two vectors is called the *dot product*.

> **The dot product of $\vec{u}$ and $\vec{v}$ is defined as**
>
> $$\vec{u} \cdot \vec{v} = (x_1,\, y_1) \cdot (x_2,\, y_2) = x_1 x_2 + y_1 y_2$$
>
> **The dot product of two vectors is not a vector — it is a <u>number</u>.**

For example: $(2,\, 1) \cdot (-3,\, 5) = 2(-3) + 1 \cdot 5 = -1$.

Notice that for a vector $\vec{u} = (x,\, y)$, $\vec{u} \cdot \vec{u}$ is equal to the squared length (magnitude) of $\vec{u}$: $|\vec{u}|^2 = (x, y) \cdot (x, y) = x^2 + y^2$.

A similar definition of dot product applies to n-dimensional vectors:

$$(u_1, u_2, ..., u_n) \cdot (v_1, v_2, ..., v_n) = u_1 v_1 + u_2 v_2 + ... + u_n v_n$$

Again, $(\vec{u} \cdot \vec{u}) = |\vec{u}|^2 = u_1^2 + u_2^2 + ... + u_n^2$.

Example 2

```
def dot_product(_u, _v):
    """Return the dot product of two n-dimensional vectors _u and _v."""
    return sum(_u[i]*_v[i] for i in range(len(_u)))
```

❖ ❖ ❖

The dot product operation is important because it helps calculate a projection of one vector onto another and the angle between two vectors. It is not too hard to prove that the dot product of vectors $\vec{u}$ and $\vec{v}$ in n-dimensional space is equal to the product of their lengths times the *cosine* of the angle between them:

> $$\vec{u} \cdot \vec{v} = |\vec{u}| \cdot |\vec{v}| \cos \theta$$

$\cos \theta$ (cosine of theta) is the trigonometric function (Figure 14-4).

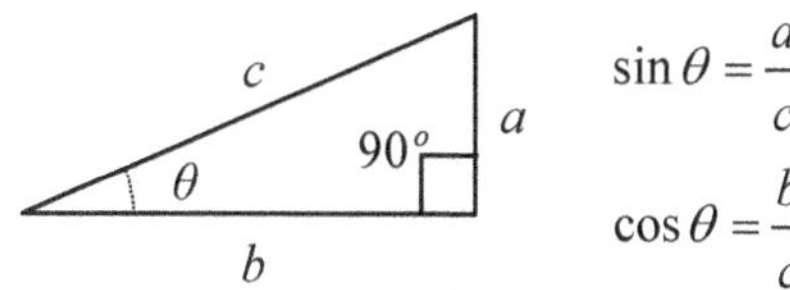

Figure 14-4. The definitions of the sin and cos functions in geometry

Python's `math` module includes the `sin` and `cos` functions and their inverse functions, `asin` and `acos`: if `y = sin(theta)` then `theta = asin(y)`, and if `y = cos(theta)` then `theta = acos(y)`. In all these functions `theta` is measured in *radians*.

$$360° = 2\pi \text{ radians, so one degree is } \frac{\pi}{180} \text{ radians.}$$

$\dfrac{\pi}{2}$ radians is 90°; $\cos\dfrac{\pi}{2} = 0$, so

the dot product of two *orthogonal* (that is, *perpendicular* to each other, forming a 90° angle) vectors is 0.

Example 3

What is the angle between $\vec{u} = (7, 2)$ and $\vec{v} = (4, 9)$?

Solution

$$\vec{u}\cdot\vec{v}=|\vec{u}|\cdot|\vec{v}|\cos\theta. \quad \vec{u}\cdot\vec{v}=46;\ \vec{u}\cdot\vec{u}=53 \ \Rightarrow\ |\vec{u}|=\sqrt{53}\,;\ \vec{v}\cdot\vec{v}=97 \ \Rightarrow\ |\vec{v}|=\sqrt{97}\,.$$

$$\cos\theta=\frac{\vec{u}\cdot\vec{v}}{|\vec{u}|\cdot|\vec{v}|}=\frac{46}{\sqrt{53}\sqrt{97}}\,.$$

```
>>> from math import sqrt, acos, degrees
>>> acos(46/(sqrt(53)*sqrt(97)))
0.8742723382105562
>>> degrees(0.8742723382105562)
50.09211512449896
```

The angle is approximately 0.874 radians, approximately 50°.

Section 14.2 ~ Exercises

1.

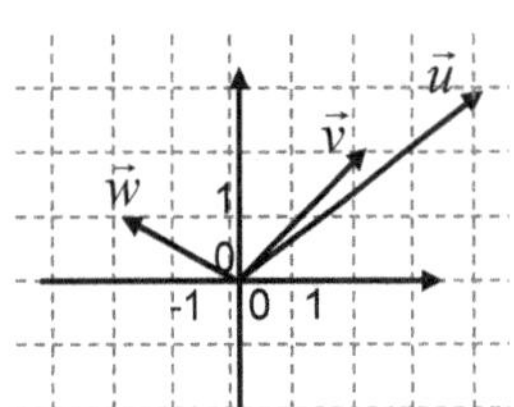

Represent each of the vectors shown above in component form (as a pair of coordinates. (One square on the grid is one unit.) What are the lengths of these vectors? ✓

2.   If $\vec{v}=(x,\,y)$, what is $-\vec{v}$?

3.   State (exactly) the lengths of the following vectors. ✓
(a) (5, 12)   (b) (1, 7)   (c) (−3, 4)   (d)■ (3, 4, 12)

4.   Find a unit vector that has the same direction as (3, −4). ✓

5.

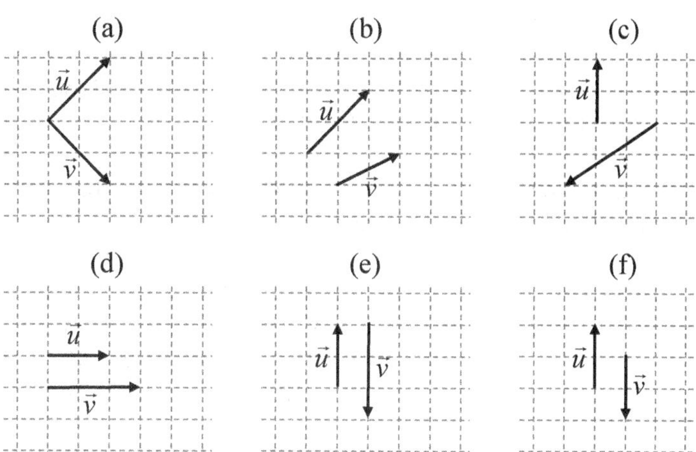

For each pair of vectors $\vec{u}$ and $\vec{v}$ above, represent each $\vec{u}$ and $\vec{v}$ in the component form and state their sum $\vec{u}+\vec{v}$ and difference $\vec{u}-\vec{v}$. ✓

6.   Determine the angle, in degrees and radians, between the following vectors, using their dot product and Python's `math` functions.

(a)   (3, 1) and (2, 5)
(b)   (3, 1) and (−2, 5)
(c)■ (3, 1, 1) and (1, −2, 5)

7.   What is the acute angle, in radians, between the lines $y = 2x$ and $y = \dfrac{1}{2}x$ on the coordinate plane? ✓

8.   Find a two-dimensional vector that is orthogonal to (1, 5) and has the same length. How many such vectors exist? ✓

9.   In Python's turtle graphics, start at the "home" position at the origin, turn by a random angle, and go forward by a random distance while drawing a line. Save the turtle's position. You have drawn a vector from the origin to the turtle's position. Go back to the origin (without changing the turtle's direction), turn left or right 90 degrees, go forward by some other random distance, and save the turtle's position. You have drawn another vector, orthogonal to the first. Calculate and display the dot product of these two vectors; it should be 0 or very close to 0.

10.   (a) Give an example of two two-dimensional unit vectors that are orthogonal to each other.

(b) Give an example of three three-dimensional unit vectors such that any two of them are orthogonal to each other.

(c)■ Give an example of n n-dimensional unit vectors such that any two of them are orthogonal to each other.

11.■   Write and test a Python function that returns the three angles (in degrees) of a triangle given by the coordinates of its vertices. What are the angles of the triangle with vertices in (0, 4), (4, 5), and (5, 2)? Are all the angles in this triangle acute (that is, less than 90°), or is one of the angles a right angle or an obtuse angle?

12.■   Prove the Schwarz inequality:

$$(u_1 v_1 + u_2 v_2 + \ldots + u_n v_n)^2 \le (u_1^2 + u_2^2 + \ldots + u_n^2)(v_1^2 + v_2^2 + \ldots + v_n^2)$$

13.◆   Suppose that $x + y + z = 1$. When $x = y = z = \dfrac{1}{3}$, $x^2 + y^2 + z^2 = \dfrac{1}{3}$. Show that $x^2 + y^2 + z^2 > \dfrac{1}{3}$ in all other cases. ⸨ Hint: See Question 12. ⸩ ✓

14.◆ $x + y + z = 1$ is the equation for the plane in three-dimensional space that contains the points (1, 0, 0), (0, 1, 0), and (0, 0, 1):

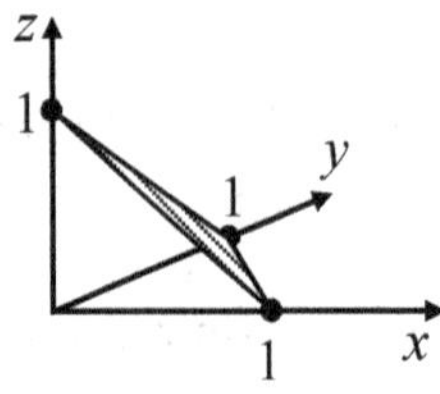

Find the shortest distance from the origin to this plane. ⸨ Hint: See Question 13. ⸩

14.3 Matrices

The mathematical term *matrix* refers to a rectangular table of numbers. Each element of a matrix can be identified by two subscripts: the row and the column number. For example:

$$A = \begin{pmatrix} a_{11} & a_{12} & a_{13} \\ a_{21} & a_{22} & a_{23} \end{pmatrix}$$

The numbers in the matrix are called its *elements*.

There is no special structure in Python for representing a matrix. Typically, each row of a matrix is represented as a list (or a tuple), and the whole matrix is represented as a list of lists (or tuples). For example,

```
1   2   3
4   5   6
```

can be defined as:

```
m = [[1, 2, 3], [4, 5, 6]]
```

Since Python lets you split lists between lines, it is better to write

```
m = [[1, 2, 3],
     [4, 5, 6]]
```

In math, the subscripts usually start from 1; you have to translate them into Python indices, which start from 0.

If `m` is a matrix, `m[0]` is its first row, `m[1]` is its second row, and so on. `m[r]` refers to the row with the index `r` (that is, the (r+1)-th row of the matrix).

The elements in the row with the index `r` are `m[r][0]`, `m[r][1]`, `m[r][2]`, and so on. In the above example, the value of the element `m[0][2]` is 3.

> **If `m` is a matrix defined as a list of its rows, `m[r][c]` refers to the element in the row with the index `r` and the column with the index `c`. `len(m)` gives the number of rows; `len(m[r])` gives the number of elements in the *r*-th row, which is the number of columns in the matrix.**

In Python, a list of lists represents a two-dimensional *array*. The rows of this array do not have to be the same length. For example, you can make a "triangular" array with one element in the first row, two elements in the second row, and so on.

Example 1

Write a Python function that returns the sum of all the elements of a given matrix.

Solution

```python
def sum_elements(m):
    """Return the sum of all the elements in the matrix m."""
    sum_m = 0
    for r in range(len(m)):
        for c in range(len(m[0])):
            sum_m += m[r][c]
    return sum_m
```

Or:

```python
def sum_elements(m):
    sum_m = 0
    for row in m:
        for a in row:
            sum_m += a
    return sum_m
```

Or, even shorter:

```python
def sum_elements(m):
    sum_m = 0
    for row in m:
        sum_m += sum(row)
    return sum_m
```

Or, even shorter:

```python
def sum_elements(m):
    return sum(sum(row) for row in m)
```

❖   ❖   ❖

A matrix in Python can hold any types of objects. For example, you can represent a chess position as an 8 by 8 matrix in which each element either holds a chess piece or `None`.

 **You need to create a matrix before you can store values in it.**

Example 2

Write a Python function that creates and returns a matrix with a given number of rows and columns, filled with a given value x.

Solution

```python
def empty_matrix(n_rows, n_cols, x=0):
    """Return an n_rows by n_cols matrix with all elements set to x."""
    m = []
    for r in range(n_rows):
        m.append(n_ncols*[x])
    return m
```

We start with an empty list of rows m and append a new row to it n_rows times. n_cols*[x] creates one row, which is a list of n_cols elements, all set to x.

❖   ❖   ❖

In the above example, you might find it wasteful to repeat n_cols*[x] several times. You might be tempted to take a shortcut: create a row once, then just append it to the matrix n_rows times, like this:

```python
def empty_matrix(n_rows, n_cols, x=0):   # buggy code!
    m = []
    row = n_cols*[x]
    for r in range(n_nows):
        m.append(row)
    return m
```

At first, everything seems OK:

```python
>>> m = empty_matrix(2, 3)
>>> m
[[0, 0, 0], [0, 0, 0]]
```

But try this:

```python
>>> m[0][0] = 1
>>> m
```

You get

```
[[1, 0, 0], [1, 0, 0]]
```

This is because the rows in your matrix actually refer to the same list!

A square matrix defines a *linear transformation* (a linear mapping or function) on n-dimensional vectors. This is the main function (pardon the pun) of matrices in linear algebra. Given an n by n matrix

$$A = \begin{pmatrix} a_{11} & a_{12} & \dots & a_{1n} \\ a_{21} & a_{22} & \dots & a_{2n} \\ \dots & & & \\ a_{n1} & a_{n2} & \dots & a_{nn} \end{pmatrix}$$

and an n-dimensional vector

$$\vec{u} = \left(u_1, u_2, \dots, u_n \right),$$

$A\vec{u}$ is a new vector $\vec{h} = \left(h_1, h_2, \dots, h_n \right)$, such that $h_i = a_{i1}u_1 + a_{i2}u_2 + \dots + a_{in}u_n$.   In other words, h_i is the dot product of the i-th row of the matrix and $\vec{u}$.

Example 3

$$A = \begin{pmatrix} 1 & 3 \\ 2 & 5 \end{pmatrix}, \ \vec{u} = (4, 7) \ \Rightarrow \ A\vec{u} = \left((1, 3) \cdot (4, 7), (2, 5) \cdot (4, 7) \right) = (25, 43).$$

> **We say that the transformation $\vec{u} \to A\vec{u}$ is *linear*, because for any vectors $\vec{u}$ and $\vec{v}$ and any number k, $A(\vec{u} + \vec{v}) = A\vec{u} + A\vec{v}$ and $A(k\vec{u}) = kA\vec{u}$.**

Suppose we have two *n* by *n* matrices:

$$A = \begin{pmatrix} a_{11} & a_{12} & \dots & a_{1n} \\ a_{21} & a_{22} & \dots & a_{2n} \\ \dots & & & \\ a_{n1} & a_{n2} & \dots & a_{nn} \end{pmatrix} \quad \text{and} \quad B = \begin{pmatrix} b_{11} & b_{12} & \dots & b_{1n} \\ b_{21} & b_{22} & \dots & b_{2n} \\ \dots & & & \\ b_{n1} & b_{n2} & \dots & b_{nn} \end{pmatrix}$$

The matrix C in which $c_{ij} = a_{i1}b_{1j} + a_{i2}b_{2j} + \dots + a_{in}b_{nj}$ is called the *product* of A and B and denoted as $A \cdot B$ or, more often, simply AB. c_{ij} is the dot product of the *i*-th row in A and the *j*-th column in B.

The product of two matrices defines a composition of two linear transformations:

$$(AB)\vec{u} = A(B\vec{u})$$

The product of three matrices obeys the associative law:

$$A(BC) = (AB)C$$

But multiplication of matrices is not commutative. For example,

$$\begin{pmatrix} 3 & 1 \\ 7 & 2 \end{pmatrix} \cdot \begin{pmatrix} 5 & 2 \\ 3 & 1 \end{pmatrix} = \begin{pmatrix} 18 & 7 \\ 41 & 16 \end{pmatrix} \quad \text{while} \quad \begin{pmatrix} 5 & 2 \\ 3 & 1 \end{pmatrix} \cdot \begin{pmatrix} 3 & 1 \\ 7 & 2 \end{pmatrix} = \begin{pmatrix} 29 & 9 \\ 16 & 5 \end{pmatrix}.$$

❖   ❖   ❖

Certain linear transformations defined by matrices preserve the lengths of all the vectors. Such a transformation just rotates all the vectors around the origin. The matrix that rotates vectors on the plane counterclockwise by the angle θ (usually measured in radians) is

$$\begin{pmatrix} \cos\theta & -\sin\theta \\ \sin\theta & \cos\theta \end{pmatrix}$$

See Questions 7 and 8 in the exercises.

❖   ❖   ❖

Matrices and vectors can be used to describe and solve systems of linear equations. This is the general form of a system of n linear equations in n variables:

$$a_{11}x_1 + a_{12}x_2 + ... + a_{1n}x_n = c_1$$
$$a_{21}x_1 + a_{22}x_2 + ... + a_{2n}x_n = c_2$$
$$\text{... ... ...}$$
$$a_{n1}x_1 + a_{n2}x_2 + ... + a_{nn}x_n = c_n$$

In matrix/vector form it can be written as $A\vec{x} = \vec{c}$, where A is the n by n matrix of the coefficients, $\vec{x} = (x_1, x_2, ..., x_n)$ and $\vec{c} = (c_1, c_2, ..., c_n)$. For two equations in two variables, we get

$$a_{11}x_1 + a_{12}x_2 = c_1$$
$$a_{21}x_1 + a_{22}x_2 = c_2$$

Or, using a, b, c, d for the coefficients, x, y for the variables, and e, f for the values on the right-hand side,

$$ax + by = e$$
$$cx + dy = f$$

Each of these two equations defines a straight line in the x-y coordinate plane. There are three possibilities:

- The two lines intersect in one point; the system of equations has a unique solution;

- The lines coincide; there are infinitely many solutions;

- The lines are parallel to each other; there are no solutions.

The necessary and sufficient condition for the system of equations $\begin{matrix} ax + by = e \\ cx + dy = f \end{matrix}$ to have a unique solution is for the *determinant* of the matrix $\begin{pmatrix} a & b \\ c & d \end{pmatrix}$ to not be equal to 0.

The determinant is defined as $ad - bc$ and denoted as $\begin{vmatrix} a & b \\ c & d \end{vmatrix}$.

If $\begin{vmatrix} a & b \\ c & d \end{vmatrix} \neq 0$, the solution of the above system of two equations is

$$x = \frac{\begin{vmatrix} e & b \\ f & d \end{vmatrix}}{\begin{vmatrix} a & b \\ c & d \end{vmatrix}}, \quad y = \frac{\begin{vmatrix} a & e \\ c & f \end{vmatrix}}{\begin{vmatrix} a & b \\ c & d \end{vmatrix}}$$

The definition of the determinant of a matrix is extended to n by n matrices, and the solution of a system of n linear equations in n variables $A\vec{x} = \vec{c}$ can be obtained in a similar manner: to find x_i, substitute the vector $\vec{c}$ for the i-th column of the matrix A, take the determinant, and divide it by the determinant of A. See Question 11 in the exercises.

Section 14.3 ~ Exercises

1.   Write and test a Python function that returns the sum of the elements on the main diagonal (upper left to lower right) of a square matrix represented as a list of lists. (In linear algebra, this value is called the *trace* of the matrix.)

2.   Write and test a function that takes a matrix and prints it neatly, as a rectangular matrix with aligned right-justified columns. Assume that all the entries in the matrix are positive integers with no more than two digits. ✓

3.   Write a program that reads a matrix of integers from a text file. Each line in the file holds integer values for the corresponding row of the matrix, separated by spaces, commas, or tabs. ⸴ Hint: For testing, use the function from Question 2 to display the resulting matrix. ⸵

4.   Write and test a function that *transposes* a given square n by n matrix and returns the resulting matrix. In the transposed matrix, all elements are flipped symmetrically over the main diagonal: rows become columns and columns become rows. For example,

$$\begin{pmatrix} 1 & 2 & 3 \\ 4 & 5 & 6 \\ 7 & 8 & 9 \end{pmatrix} \Rightarrow \begin{pmatrix} 1 & 4 & 7 \\ 2 & 5 & 8 \\ 3 & 6 & 9 \end{pmatrix}$$

5. Write and test a function that takes an *n* by *n* matrix *A* and an *n*-dimensional vector $\vec{u}$ as arguments and returns the vector $A\vec{u}$. ✓

6.■ Write and test a function that takes two *n* by *n* matrices and returns their product.

7.◆ Create a text file that holds the coordinates of the vertices of a polygon, one *x-y* pair per line, with *x* and *y* separated by spaces and/or commas. Write a program that reads the file and creates a polygon represented as a list of its vertices (*x-y* pairs). Your program should then display the polygon in turtle graphics *n* times, each time rotating the polygon around the origin by $\dfrac{2\pi}{n}$

radians. Use several colors in alternation or use random colors. The resulting display might look like this (for 18 rotations of a hexagon by 20 degrees each, outlined or filled):

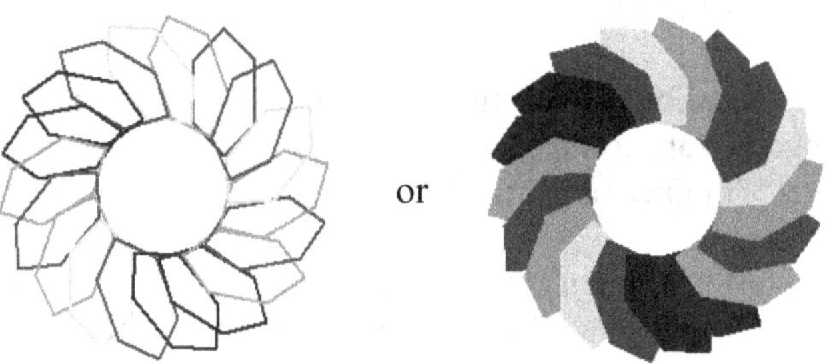

or

≷ Hints:

1. `[int(w) for w in line.replace(',', ' ').split()]`

   converts `line` into `[x, y]`.

2. Use absolute coordinates and `turtle`'s `goto` (or `setposition`) function to draw the polygon.

3. Use a rotation matrix to rotate each vertex vector around the origin.

8.   Prove that when a two-dimensional rotation matrix

$$\begin{pmatrix} \cos\theta & -\sin\theta \\ \sin\theta & \cos\theta \end{pmatrix}$$

is applied to a vector, the length of the vector remains the same. ⸨ Hint: $(\sin\theta)^2 + (\cos\theta)^2 = 1$ (see Figure 14-4 on Page 281). ⸩ ✓

9.◆   (a) Compute the determinant of the matrix of coefficients for the system of two linear equations

$$2x - 7y = -11$$
$$3x - 4y = 3$$

(b) Find the solution (x, y) using determinants.

10.◆   (a) Write a Python function that returns the determinant of a given 2 by 2 matrix.

(b) Write a function that takes a 2 by 2 matrix A and a two-dimensional vector $\vec{c}$ as parameters and attempts to solve the system of two linear equations $A\vec{x} = \vec{c}$, using determinants. If the solution exists and is unique, the function should return the vector $\vec{x}$; otherwise the function should return `(None, None)`.

11.◆   Given that

$$\begin{vmatrix} a_{11} & a_{12} & a_{13} \\ a_{21} & a_{22} & a_{23} \\ a_{31} & a_{32} & a_{33} \end{vmatrix} = a_{11}a_{22}a_{33} + a_{12}a_{23}a_{31} + a_{21}a_{32}a_{13} - a_{13}a_{22}a_{31} - a_{12}a_{21}a_{33} - a_{11}a_{23}a_{32}$$

write a function that solves a system of three linear equations in three variables. ✓

14.4 Review

Terms and notation introduced in this chapter:

Vector	*Matrix applied to a vector*	$\vec{x} \cdot \vec{y}$
Magnitude and direction	*Product of matrices*	
of a vector	*Linear transformation*	$A\vec{x}$
Dot product of vectors	*System of linear equations*	
Matrix	*Determinant*	$C = AB$
Element of a matrix		

Some of the Python features mentioned in this chapter:

```python
m = [[1, 2, 3],
     [4, 5, 6]]
n_rows = len(m)
n_cols = len(m[0])
x = m[i][j]

dot_product = sum(u[i]*v[i] for i in range(len(u))

m = []
for r in range(n_rows):
    m.append(n_cols*[x])

sum_elements = sum(sum(row) for row in m)
```

$$cx^6 + hx^5 + ax^4 + px^3 + tx^2 + ex + r = 15$$

Polynomials

15.1 Prologue

An expression in this form

$$a_n x^n + a_{n-1} x^{n-1} + \ldots + a_1 x + a_0$$

is called a *polynomial*. The numbers a_n, a_{n-1}, ..., a_0 are called the *coefficients* of the polynomial. $a_n \neq 0$ is the *leading coefficient*; a_0 is called the *constant term*. n is called the *degree of the polynomial*. We will work with polynomials whose coefficients are real numbers. For example, the expression $2.4x^3 + 3x^2 + 1.5x + 4.9$ is a polynomial of degree 3, with the leading coefficient 2.4 and the constant term 4.9.

When we write polynomials, we usually omit the terms with zero coefficients. For example, $4x^2 + 0x + 1$ is written simply as $4x^2 + 1$. Also, when the coefficient is 1 we do not write it. For example, $1x^2 + 1x + 1$ is written simply as $x^2 + x + 1$. When a polynomial has a term with a negative coefficient, that coefficient is usually written with a minus sign in place of plus and without parentheses. For example, $x^7 + (-3)x^2$ is written simply as $x^7 - 3x^2$.

In computer programs, a polynomial can be represented as a list or tuple of its coefficients. For example, in Python $1.8x^5 - 3x^2 + 2.35$ can be represented as p = [1.8, 0, 0, -3, 0, 2.35] or p = (1.8, 0, 0, -3, 0, 2.35). An n-th degree polynomial is represented by a list or tuple of length $n+1$.

There are two ways to look at polynomials. First, we can view a polynomial as a function $P(x)$ defined by the formula $P(x) = a_n x^n + a_{n-1} x^{n-1} + \ldots + a_1 x + a_0$. We can calculate the value of $P(x)$ for any real number x. Second, we can view polynomials as abstract algebraic objects. We can add, subtract, multiply, and factor polynomials. We can study properties of polynomials with integer, rational, or real coefficients. In many important ways, the algebraic properties of polynomials parallel some properties of integers. For example, any positive integer can be factored, or represented as a product of primes, and such a representation is unique. Similarly, any polynomial can be uniquely represented as a product of "prime" polynomials (give or take constant factors). For example, $x^3 - 1 = (x-1)(x^2 + x + 1)$.

$a_2x^2 + a_1x + a_0$ is a second-degree polynomial. It is also called a *quadratic polynomial* and it defines a *quadratic function*. The letters *a*, *b*, and *c* are often used for the coefficients when we deal with the general form of a quadratic polynomial:

$$ax^2 + bx + c$$

$ax + b$, a first-degree polynomial, is called a *linear polynomial* or *linear function*. The polynomial of zero degree is simply a constant. It is convenient to consider 0 as a polynomial of degree 0.

Polynomials are important in mathematics and computer technology because they can be used to approximate other functions and smooth curves. For example, in scalable ("TrueType") fonts on your computer, the contour of each character is composed of quadratic *splines* (pieces of quadratic curves). It is easy to compute the value of a polynomial for a given *x*. Polynomials also serve as interesting models in abstract algebra.

15.2 Addition and Subtraction

The sum of two polynomials is a polynomial.

$$\left(a_n x^n + \ldots + a_1 x + a_0\right) + \left(b_n x^n + \ldots + b_1 x + b_0\right) =$$
$$\left(a_n + b_n\right)x^n + \ldots + \left(a_1 + b_1\right)x + \left(a_0 + b_0\right)$$

In other words, to add two polynomials we simply add the respective coefficients. If the degree of one polynomial is lower than the degree of the other, the missing terms in the lower-degree polynomial are considered to have zero coefficients.

Example 1

$$(3x^3 + 2x^2 + 7x + 2) + (4x^2 + 3x + 6) = 3x^3 + 6x^2 + 10x + 8$$

$$\begin{array}{r} 3x^3 + 2x^2 + 7x + 2 \\ + 4x^2 + 3x + 6 \\ \hline 3x^3 + 6x^2 + 10x + 8 \end{array}$$

> **It is important to combine properly the *like terms* (that is, the terms with the same degree of x) when adding two polynomials.**

Example 2

$$(3x^3 + 2x^2 + 7x + 2) + (4x^2 + 1) = 3x^3 + 6x^2 + 7x + 3$$

$$
\begin{array}{r}
3x^3 + 2x^2 + 7x + 2 \\
+\ \ \ \ \ \ \ \ 4x^2\ \ \ \ \ \ + 1 \\
\hline
3x^3 + 6x^2 + 7x + 3
\end{array}
$$

With a little practice, you can quickly learn to combine the like terms in your head and write the resulting sum without writing down the intermediate steps.

Example 3

$$(3x^3 - 2x^2 - 7x + 2) + (4x^2 - 1) =$$
$$3x^3 + (-2 + 4)x^2 - 7x + (2 - 1) =$$
$$3x^3 + 2x^2 - 7x + 1$$

It is better to perform all the intermediate steps quickly in your head and write only the final result. Something like this: "for x cubed: 3 and nothing ==> $3x^3$; for x squared: $-2 + 4$ ==> $+2x^2$; for x: $-7 +$ nothing ==> $-7x$; for constant terms: $+2 - 1$ ==> $+1$."

❖   ❖   ❖

To negate a polynomial you need to negate all of its coefficients. $P_1 - P_2$ is the same as $P_1 + \left(-P_2\right)$.

Example 4

$$(x^2 - 1) - (2x^3 + x^2 - 5x + 1) =$$
$$(x^2 - 1) + (-2x^3 - x^2 + 5x - 1) =$$
$$-2x^3 + 5x - 2$$

Section 15.2 ~ Exercises

1.   Simplify $\left(x^4 + 3x^2 + 1\right) + \left(x^3 + x^2 + 8\right)$. ✓

2.   Simplify $\left(2x^4 + x^3 + 4\right) - \left(x^4 + 2x^3 + 5\right)$.

3.   Simplify $\left(2.5x^4 - 1.2x^3 + 3\right) - \left(-2x^3 + 1\right)$.

4.   Write and test a Python function `negate(p)` that negates the polynomial represented by the list of its coefficients `p` and returns a new polynomial (represented as a list). ✓

5.   Write and test a function `add(p1, p2)` that adds the polynomials `p1` and `p2` (represented as lists) and returns their sum (represented as a list). When you add or subtract polynomials, several leading coefficients in the result may become zeros. Don't forget to get rid of them.

⊸ Hints:

1. The degree of the sum does not exceed the largest of the degrees of `p1` and `p2`.

2. Find the first non-zero element of the resulting list and return the slice of the list starting from that element.

6.   Write and test a Python function `eval_polynomial(p, x)` that returns the value of $P(x)$, where P is the polynomial represented by the list of its coefficients `p`. For example, `eval_polynomial([1, 0, 3], 2)` should return $1 \cdot 2^2 + 0 \cdot 2 + 3 = 7$. Use a single `while` loop. ⊸ Hint: keep track of x^k. ⊸

7. ▪   Write and test a version of `eval_polynomial(p, x)`, based on the
formula $a_n x^n + a_{n-1} x^{n-1} + ... + a_1 x + a_0 = (...((a_n)x + a_{n-1})x + ... + a_1)x + a_0$.
Use one `while` loop. ⸮ Hint: start with $y = a_n$ and go from there:
$y \leftarrow y \cdot x + a_{n-1}$, etc. ⸮

8.   Using a graphing calculator or a function plotting app, plot on the same
screen $y = \sin(x)$ and $y = x - \dfrac{x^3}{6}$. For which values of x does the
polynomial $x - \dfrac{x^3}{6}$ give a reasonable approximation of $\sin(x)$? ✓

15.3 Multiplication, Division, and Roots

To multiply a polynomial by a number, we multiply each coefficient by that number.
To multiply a polynomial by x^k, we add k to each exponent.

Example 1

$$\left(3x^3 - x^2 + 2x - 6\right) \cdot 3x^2 = 9x^5 - 3x^4 + 6x^3 - 18x^2$$

❖   ❖   ❖

If a polynomial is represented in a Python program as a list of its coefficients, then
multiplication by x^k is equivalent to padding the list with k zeros at the end:

```
p += k*[0]
```

Example 2

$x^2 - 5x + 6$ is represented as `[1, -5, 6]`. What list represents $x^3 \cdot \left(x^2 - 5x + 6\right)$?

Solution

```
[1, -5, 6, 0, 0, 0]
```

❖   ❖   ❖

To multiply a polynomial by another polynomial , we multiply each term of the first operand by each term of the second operand, then collect the like terms (with the appropriate signs).

Example 3

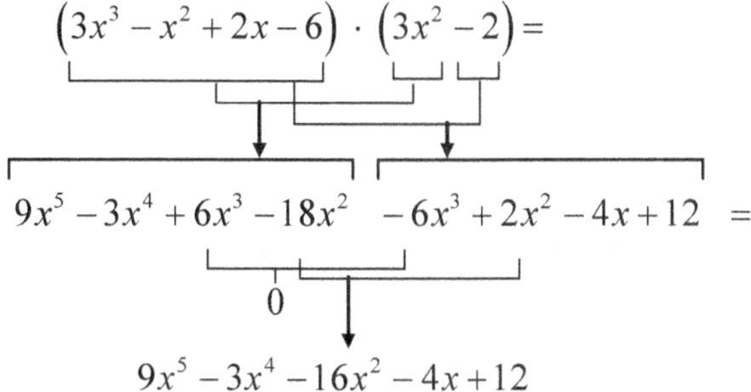

$$9x^5 - 3x^4 - 16x^2 - 4x + 12$$

The number of terms in the product, before collecting the like terms, is equal to the number of terms in the first polynomial times the number of terms in the second polynomial (here $4 \cdot 2 = 8$).

❖ ❖ ❖

Many algebra textbooks teach the so-called "FOIL" (First, Outer, Inner, Last) method and mnemonic rule for multiplying two linear polynomials. For example, $(2x-1)(3x+5) = 6x^2 + 10x - 3x - 5$. FOIL works only when multiplying two linear polynomials. It may mislead you when one or both factors are quadratics or polynomials of higher degrees. To find $P \cdot Q$, it is better to multiply each term of P by the first term of Q, then each term of P by the second term of Q, and so on. In the above example, we would get $(2x-1)(3x+5) = 6x^2 - 3x + 10x - 5$. This is not a big deal for linear polynomials — one can get used to either order — but things would be clearer if "FOIL" didn't exist, and you were taught the more general method from the outset. In either case, you have to collect the like terms at the end:

$$(2x-1)(3x+5) = 6x^2 - 3x + 10x - 5 = 6x^2 + 7x - 5$$

❖ ❖ ❖

Given two positive integers, we can divide one (the dividend) by the other (the divisor) and get a quotient and a remainder. For example, when we divide 17 by 3, the quotient is 5 and the remainder is 2: $17 = 5 \cdot 3 + 2$. The remainder is smaller than the divisor: $2 < 3$. A similar type of division with a remainder can be performed on polynomials with real coefficients. For example, if we divide $x^5 - 6.9x^4 + 5.3x^3 - 4x^2 + 3x + 1.2$ by $2x^2 - x + 0.2$, the quotient is $0.5x^3 - 3.2x^2 + x - 1.18$ and the remainder is $1.62x + 1.436$:

$$x^5 - 6.9x^4 + 5.3x^3 - 4x^2 + 3x + 1.2 =$$
$$\left(0.5x^3 - 3.2x^2 + x - 1.18\right)\left(2x^2 - x + 0.2\right) + \left(1.62x + 1.436\right)$$

The <u>degree</u> of the remainder is smaller than the <u>degree</u> of the divisor.

We can use the same *long division* algorithm for polynomials as we use for integers (Figure 15-1). As you can see, this can get quite tedious. Fortunately, CAS (Computer Algebra System) software and calculators can perform such tasks for you. Soon you will write your own Python function that implements the long division algorithm for polynomials (see Question 5).

$$
\begin{array}{r}
0.5x^3 - 3.2x^2 + x - 1.18 \\
2x^2 - x + 0.2 \,\overline{)\, x^5 - 6.9x^4 + 5.3x^3 - 4x^2 + 3x + 1.2} \\
\underline{x^5 - 0.5x^4 + 0.1x^3} \\
-6.4x^4 + 5.2x^3 - 4x^2 + 3x + 1.2 \\
\underline{-6.4x^4 + 3.2x^3 - 0.64x^2} \\
2x^3 - 3.36x^2 + 3x + 1.2 \\
\underline{2x^3 - x^2 + 0.2x} \\
-2.36x^2 + 2.8x + 1.2 \\
\underline{-2.36x^2 + 1.18x - 0.236} \\
1.62x + 1.436
\end{array}
$$

Figure 15-1. Example of long division for polynomials

❖   ❖   ❖

When you divide a polynomial by $(x - c)$, the remainder is a zero-degree polynomial, which is simply a constant:

$$P(x) = Q(x)(x - c) + r$$

r is the same for any x. It is easy to find the remainder r without division: if you plug c into the above equation for x, then $(x-c)$ becomes zero, and you get $P(c)=r$. This fact is known as the *remainder theorem*: $P(x)=Q(x)(x-c)+P(c)$.

❖ ❖ ❖

A value of x such that $P(x)=0$ is called a *root* (or a *zero*) of the polynomial $P(x)$.

> **To find all the roots of a polynomial is the same as to solve the equation $P(x)=0$.**

As you know, a linear polynomial $ax+b$ has one root, $x=\dfrac{-b}{a}$. A quadratic polynomial ax^2+bx+c can have a maximum of two real roots: $x_1=\dfrac{-b-\sqrt{b^2-4ac}}{2a}$

and $x_2=\dfrac{-b+\sqrt{b^2-4ac}}{2a}$ (when $b^2-4ac\geq 0$). In general, an n-th degree polynomial with real coefficients can have up to n real roots.

The remainder theorem states that $P(x)=Q(x)(x-c)+P(c)$. In particular, $(x-c)$ evenly divides $P(x)$ if and only if $P(c)=0$. This fact is a *corollary* to (follows directly from) the remainder theorem.

The above fact is also known as the *factor theorem*: c is a root of a polynomial $P(x)$ if and only if $(x-c)$ is a factor of $P(x)$. However, when we say "a factor of $P(x)$," we basically assume that a polynomial with real coefficients can be represented as a product of prime polynomial factors, and that such a representation is unique (up to constant factors). This is indeed true; moreover, the prime factors are all linear or quadratic polynomials. For example, $x^4+1=(x^2+\sqrt{2}x+1)(x^2-\sqrt{2}x+1)$.

The proof is not easy, though; it uses *complex numbers* and follows from the *fundamental theorem of algebra*: any n-th degree polynomial ($n>1$) with complex coefficients can be represented as a product of n linear factors (with complex coefficients) and, therefore, has exactly n complex roots (not necessarily all different). For example, $x^2+1=(x-i)(x+i)$, where i is the *imaginary number* $i=\sqrt{-1}$. It is also true that any <u>odd</u> degree polynomial with real coefficients has at least one real root.

> If $P(x)$ is an *n*-th degree polynomial $a_n x^n + ... + a_1 x + a_0$ with real
> coefficients and it has *n* roots $c_1, c_2, ..., c_n$, then
> $$P(x) = a_n (x - c_1)(x - c_2)...(x - c_n).$$

Section 15.3 ~ Exercises

1. When solving an equation, it is often convenient to divide all the coefficients
    of a polynomial by the same number equal to the leading coefficient, so that
    the leading coefficient becomes 1. Write and test a function that does this.
    ⸮ Hint: it is easier to multiply the remaining coefficients by the reciprocal of
    the leading coefficient and then set the leading coefficient to 1. ⸮ ✓

2. Write and test a function `multiply_by_one_term(p, a, k)` that
    multiplies a given polynomial `p`, represented by a list of its coefficients, by
    ax^k and returns the product as a new list.

3.■ Write and test a function `multiply(p1, p2)` that returns the product of
    two polynomials. Use one loop; within it, call `multiply_by_one_term`
    from Question 2 and `add(p1, p2)` from Question 5 in Section 15.2.

4.■ Write a more efficient version of `multiply(p1, p2)` that does not call
    other functions. Start with a list `product` of the appropriate length, filled
    with zeros. Add `p1[i]*p2[j]` to `product[i+j]` for all appropriate `i` and
    `j`. ✓

5.◆ Write and test a function `divide(p1, p2)` that divides the polynomial `p1`
    by the polynomial `p2` and returns the quotient and the remainder (combined
    into a tuple). Use the long division algorithm.

6.■ Write a version of `eval_polynomial(p, x)` that calculates the value of
    $P(x)$ using the `divide` function from Question 5 and the remainder
    theorem.

7.■   Write a program that prompts the user to enter a positive integer n and prints out the coefficients of the expansion of $(x+1)^n$ and their sum. ✓

8.■   One of the roots of the polynomial $P(x) = x^3 - 4x^2 - x + 12$ is 3. Find the other two roots. ≷ Hint: $(x-3)$ divides $P(x)$; use long division (or the Python function that you wrote in Question 5) to find the quotient. ≷

9.■   A quadratic polynomial $p(x) = 2x^2 + bx + c$ has the roots –2 and 6. Find b and c. ≷ Hint: Distribute $2(x - x_1)(x - x_2)$, where x_1 and x_2 are the roots of p, and compare the coefficients of the result to the coefficients of p. ≷

10.■   Given $u + v = p$, $uv = q$, find u and v in terms of p and q. ≷ Hint: write a quadratic equation such that u and v are its roots. ≷ ✓

15.4 Binomial Coefficients

If you expand $(x+1)^n$, you get a polynomial. What are its coefficients? You had a glimpse of them in Section 12.5 and in Question 7 in the previous section: the coefficient at x^k turns out to be $\binom{n}{k}$ (n-choose-k). In general, if you expand $(x + y)^n$, you get the sum of the terms $a_k x^k y^{n-k}$. Again, $a_k = \binom{n}{k}$. $(x+y)$ is a *binomial*, and the fact that the coefficients of $(x + y)^n$ are the numbers n-choose-k is called the *binomial theorem*.

The binomial theorem states that

$$(x + y)^n = \binom{n}{0}x^n + \binom{n}{1}x^{n-1}y + \dots + \binom{n}{n}y^n = \sum_{k=0}^{n}\binom{n}{k}x^{n-k}y^k$$

That is why the n-choose-k numbers are often called *binomial coefficients*. The formula was discovered by Isaac Newton and announced by him in 1676.

The proof of the binomial theorem goes like this:

$$(x + y)^n = \underbrace{(x + y) \cdot (x + y) \cdot \dots \cdot (x + y)}_{n}$$

When we distribute we form products of x's and y's:

$$\underbrace{x \cdot y \cdot x \cdot x \cdot \ldots \cdot y \cdot x}_{n} = x^{n-k} y^{k}$$

We pick either x or y from each set of parentheses and multiply them. When we collect the like terms, we get the coefficient for the term $x^{n-k} y^{k}$. So the question is: How many ways are there to form $x^{n-k} y^{k}$? In order to get such a term we must take y k times and x $(n-k)$ times. So the coefficient is equal to the number of ways in which we can choose k objects (here, the sets of parentheses where we picked y) from n (all sets of parentheses).

The fact that the n-choose-k numbers are also binomial coefficients is useful for establishing many of their amazing properties (see Questions 2 - 6).

Example 1

Expand $(x + y)^{4}$.

Solution

$$(x + y)^{4} = x^{4} + 4x^{3}y + 6x^{2}y^{2} + 4xy^{3} + y^{4}$$

Example 2

What is the coefficient at $x^{8}y^{10}$ in the expansion of $(x + y)^{18}$?

Solution

$$\binom{18}{10} = \frac{18!}{10! \cdot 8!} = 43758$$

It is very useful to remember the following identities, which, of course, are special cases of the binomial theorem:

$$(x + y)^2 = x^2 + 2xy + y^2 \qquad (x - y)^2 = x^2 - 2xy + y^2$$

$$(x + y)^3 = x^3 + 3x^2y + 3xy^2 + y^3 \qquad (x - y)^3 = x^3 - 3x^2y + 3xy^2 - y^3$$

❖   ❖   ❖

We can neatly arrange the n-choose-k numbers into a triangular table:

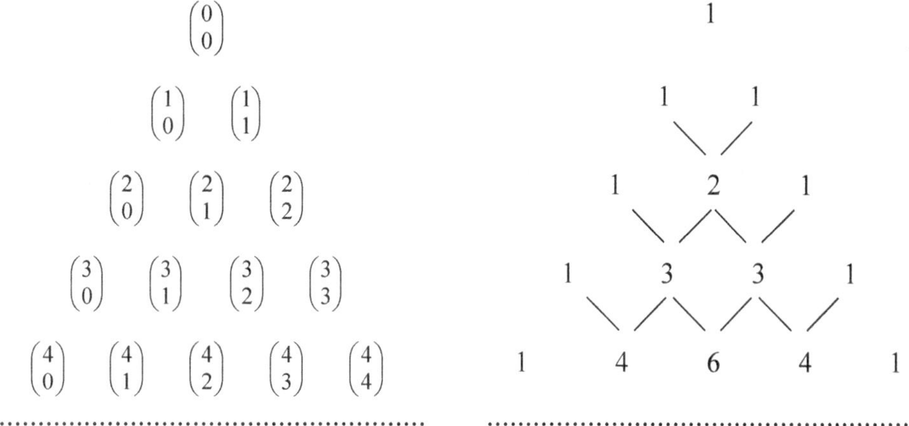

Look at this triangle. It is symmetrical about its vertical axis. The numbers on the left and right borders are 1. Each number inside is the sum of the two numbers above it: $\binom{n}{k} = \binom{n-1}{k-1} + \binom{n-1}{k}$. Starting from the top, we can build as many rows of the triangle as we need.

This triangular table is known as *Pascal's Triangle* in honor of the French mathematician and philosopher Blaise Pascal (1623-1662). According to a brief historical note in Donald Knuth's *The Art of Computer Programming*, Pascal published the table in 1653 in his *Traité du Triangle Arithmétique*. This was one of the first works on probability theory. However, Pascal was not the first to describe the n-choose-k numbers or the triangle. According to Knuth, the triangle appeared in the treatise "The Precious Mirror of the Four Elements" by the Chinese mathematician Shih-Chieh Chu in 1303, where it was said to be old knowledge. In about 1150, the Indian mathematician Bhāskara gave a very clear explanation of the n-choose-k numbers.

There is a simple combinatorial proof of the relationship $\binom{n}{k} = \binom{n-1}{k-1} + \binom{n-1}{k}$.

Suppose you have n objects and wish to choose k of them. Call one of them x and set it aside. Any combination of k objects out of n either includes x or it doesn't. The number of the ones that include x is $\binom{n-1}{k-1}$ (we need to choose the remaining $k-1$ out of $n-1$). The number of the ones that do not include x is $\binom{n-1}{k}$. The sum of these two terms is the total number of ways to choose k objects from n objects.

Section 15.4 ~ Exercises

1. Write a program that prompts the user for a positive number n and prints out a list of the binomial coefficients $\binom{n}{0}, \binom{n}{1}, ..., \binom{n}{n}$. ⁊ Hint: see Question 4 in Section 12.5. ⁊

2. If a set A has n elements, the total number of its subsets (including the empty set and the whole set A) is $\binom{n}{0} + \binom{n}{1} + ... + \binom{n}{n}$. Using this fact, derive a formula for this sum. ✓

3. Derive the formula for $\binom{n}{0} + \binom{n}{1} + ... + \binom{n}{n}$ (see Question 2) in a different way, using the binomial theorem. ⁊ Hint: consider $(1+1)^n$. ⁊

4. Show that

$$\sum_{k=0}^{n} 2^k \binom{n}{k} = \binom{n}{0} + 2\binom{n}{1} + 4\binom{n}{2} + ... + 2^n \binom{n}{n} = 3^n. \checkmark$$

5. Show that

$$\sum_{k=0}^{n} (-1)^k \binom{n}{k} = \binom{n}{0} - \binom{n}{1} + \binom{n}{2} - ... + (-1)^n \binom{n}{n} = 0$$

6. ■   Show that

$$\binom{n}{0}^2 + \binom{n}{1}^2 + \dots + \binom{n}{n}^2 = \binom{2n}{n}$$

≋ Hint: $(x+y)^{2n}$ is the same polynomial as $(x+y)^n(x+y)^n$. Compare the coefficients at $x^n y^n$ when you expand each of these representations. ≋

7. ◆   Write and test a function that generates and returns the first n rows of Pascal's triangle. The triangle is represented as a list of lists: the first list has one element, the second list has two elements, and so on. Do not multiply any polynomials in this program — just build the next row of the triangle from the preceding row using the relationship $\binom{n}{k} = \binom{n-1}{k-1} + \binom{n-1}{k}$. ✓

8. ■   Show that for any $n \geq 1$, $\binom{2n}{n} \leq \binom{2n}{n-1} + \binom{2n}{n+1}$ ≋ Hint: use Pascal's Triangle. ≋ ✓

9. ◆   Let's slice Pascal's Triangle into diagonal bands as shown below:

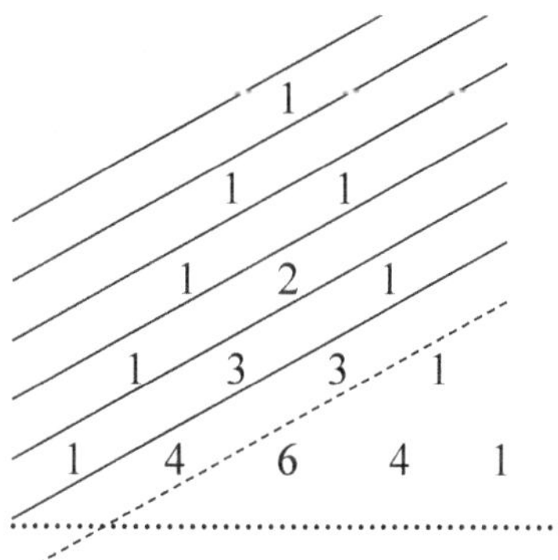

Let's take the sum of the binomial coefficients in each band. Guess what sequence these sums form and show that your guess is correct.

15.5 Review

Terms and formulas introduced in this chapter:

Polynomial
Coefficient of a polynomial
Degree of a polynomial
Linear polynomial
Quadratic polynomial
Long division (for polynomials)
Remainder theorem
Factor theorem
Binomial coefficient
Binomial theorem
Pascal's Triangle

$$P(x) = Q(x)(x-c) + P(c)$$

c is a root of $P(x)$ if and only if $(x-c)$ is a factor of $P(x)$.

If $a_n x^n + ... + a_1 x + a_0$ has n roots $c_1, c_2, ..., c_n$, then
$$a_n x^n + ... + a_1 x + a_0 = a_n (x - c_1)(x - c_2)...(x - c_n).$$
$$(x + y)^2 = x^2 + 2xy + y^2$$
$$(x - y)^2 = x^2 - 2xy + y^2$$

$$(x + y)^3 = x^3 + 3x^2 y + 3xy^2 + y^3$$
$$(x - y)^3 = x^3 - 3x^2 y + 3xy^2 - y^3$$

$$(x + y)^n = \binom{n}{0} x^n + \binom{n}{1} x^{n-1} y + ... + \binom{n}{n} y^n = \sum_{k=0}^{n} \binom{n}{k} x^{n-k} y^k$$

```
>>> def chapter(n):
        return 1 if n is 1 else chapter(n-1) + 1

>>> chapter(16)
16
```

Recurrence Relations and Recursion

16.1 Prologue

In Chapter 3 we talked about different ways to define a function: in words, with a table, with a chart, or with a formula. But we left out one very important method: we can also describe a function, especially a function on positive (or non-negative) integers, *recursively*. Here is an example. Suppose you state that a function f with the domain of all positive integers is defined as follows:

$$f(n) = \begin{cases} 1, \text{ if } n = 1 \\ n \cdot f(n-1), \text{ if } n > 1 \end{cases}$$

This definition does not tell us right away how to find the value of $f(n)$. $f(n)$ is described in terms of $f(n-1)$ (except for one simple *base case*, $n = 1$). And yet, with some work, we can find the value of $f(n)$ for any positive n. Let's see: we know that $f(1) = 1$. Next: $f(2) = 2 \cdot f(1) = 2 \cdot 1 = 2$. Next: $f(3) = 3 \cdot f(2) = 3 \cdot 2 = 6$. Next: $f(4) = 4 \cdot f(3) = 4 \cdot 6 = 24$. And so on. There is only one function that satisfies the above definition. In this case you can easily find a formula to describe this function: it is our old friend $f(n) = n!$ (*n*-factorial).

Recursion is a powerful tool for defining functions. The designers of computer hardware and programming languages have made sure that recursion is also supported in programming languages.

16.2 Recurrence Relations

Recall that a function defined on the set of all positive integers can be expressed as a sequence of the function's values: $a_1, a_2, ..., a_n, ...$ To define such a function recursively, we define each term of the sequence, except the first (or, perhaps, the first few), through its relation to the previous term(s). The relation that connects the *n*-th term to one or more of the previous terms is often the same for each n. It is called a *recurrence relation*. For example:

$$\begin{cases} a_1 = 1 \\ a_n = na_{n-1}, \text{ for any } n \geq 2 \end{cases}$$

We get, again, $a_n = n!$. This is the same as our recursive definition of $n!$, just rewritten in a slightly different way.

With a computer, it is easy to calculate the values of a sequence defined through a recurrence relation. Recall, for example, our calculation of $s(n) = 1 + 2 + ... + n$ in Chapter 4. We noticed that $s(0) = 0$ and $s(n) = s(n-1) + n$, for any $n \geq 2$. This led to the following Python code:

```python
n_max = 10
sum_n = 0
for n in range(1, n_max+1):
    sum_n += n
    print(n, sum_n)
```

Here `sum_n` stands for $s(n)$. Iterations through the `for` loop are equivalent to calculating successive values of $s(n)$.

Section 16.2 ~ Exercises

1. Consider a function

$$f(n) = \begin{cases} 1, \text{ if } n = 1 \\ 2 \cdot f(n-1) + 1, \text{ if } n > 1 \end{cases}$$

What is $f(4)$? ✓

2. A function on positive integers is described by

$$f(n) = \begin{cases} 1, \text{ if } n = 1 \\ f(n-1) + 2, \text{ if } n > 1 \end{cases}$$

Redefine it by a simple non-recursive formula.

3. Define $f(n) = 2^n$ recursively, in terms of $f(n-1)$, for all integers $n \geq 1$. ✓

4. Write a recursive definition of a function $f(n)$ that has the following properties: $f(1) = 10$, $f(2) = 30$, and the values $f(1)$, $f(2)$, ..., $f(n)$, ... form an arithmetic sequence. ✓

5. The same as Question 4, but for a geometric sequence.

6.▪ Describe the function from Question 1 with a simple *closed-form* formula (without recursion).

7.▪ Define recursively, without factorials, the function $f(n) = \dfrac{n!}{(n-3)!}$ for integers $n \geq 3$. ✓

8. Given $a_1 = 1$, $a_n = a_{n-1} + n$ for any $n \geq 2$, describe a_n with a simple *closed-form* (non-recursive) formula in terms of n.

9.◆ Find a sequence that satisfies the equation $a_n = 8a_{n-1} - 15a_{n-2}$, if $n > 2$ and, at the same time, is a geometric sequence with $a_1 = 1$. How many such sequences are there?

16.3   Recursion in Programs

Now back to our favorite, $s(n) = 1 + 2 + \ldots + n$. As we have seen, the recursive definition of $s(n)$ is $s(n) = \begin{cases} 1, \text{ if } n = 1 \\ s(n-1) + n, \text{ if } n > 1 \end{cases}$. Now suppose, just out of curiosity, that we implement this definition directly in a Python function:

```python
def add_numbers(n):
    '''Return 1 + 2 + ... + n.'''
    if n == 1:
        return 1
    else:
        return add_numbers(n-1) + n
```

Will this work? What will `add_numbers(5)` return? As you know, in a program, a function translates into a piece of code that is callable from different places in the program. The caller passes to the function some argument values and a return address — that is, the address of the instruction to which to return control after the function finishes its work. So how can this work when a piece of code passes control to itself? Won't the program get confused? Won't Python choke trying to swallow its own tail?

And yet it works! Try it:

```
>>> add_numbers(5)
15
```

Because recursion is such a useful tool in programming, the developers of Python (and other programming languages) have made special efforts to ensure that it works. There are also CPU instructions that facilitate implementing recursive calls.

Here is an analogy of what happens. Suppose one day you receive a text from a friend: What's $1+2+...+100$? They really need to know, as if their life depended on it. But you have no clue. You need to text and ask someone really smart. Gauss[*] would know, but he doesn't have a smartphone and anyway he's been dead for 150 years. So where do you find someone smart? As it so often happens, the smartest person around is you! (But can you really text yourself? We've tried, and it works!) But if you ask yourself the same question, you'll just go in circles and get nowhere.

Then you get an idea: What if I ask an easier question: What's $1+2+...+99$? Then I can just add 100 to the answer and be done. So you text yourself again. Unfortunately, you still have no clue. So you text yourself an even easier question — What's $1+2+...+98$? — hoping that if you get an answer to that, you'll just add 99 and then 100 to it. Still no clue. You keep texting yourself easier and easier questions, until finally you receive: What's $1+2$? Ah, I know that: 3. By now you have a whole stack of texts from yourself. You happily respond to the last one: $1+2$ is 3. Then you add 3 to that and get a total of 6, which means that $1+2+3$ is 6. Then you add 4 and get a total of 10, which means that $1+2+3+4$ is 10. And so on.

Eventually you find that $1+2+...+99$ is 4950. You add 100, get 5050, and reply to your friend: $1+2+...+100$ is 5050. Now your friend will be sure that you're the smartest person around! (Hopefully your mobile plan includes unlimited texting, or your parents may disagree.)

Notice that you never got confused with all the texts because they are neatly arranged in a stack and you were handling them in a *last-in-first-out* (LIFO) manner.

---

[*] Carl Friedrich Gauss (1777-1855), a famous German mathematician. Legend has it that he was asked this question by his elementary school teacher, who was trying to keep Gauss busy. It didn't work.

This seems like a lot of texting, but something like this is easy for a computer. In Python, your activity may be represented precisely as the recursive function `add_numbers` above.

Python allocates a special chunk of memory, called a *stack*, for recursive function calls. The CPU has a *push* instruction that puts a data item onto the top of the stack. The complementary *pop* instruction removes and returns the top item. (Python's list methods `append` and `pop` work in a similar way.) A function holds all its parameters, local variables, and the return address on the stack; when it calls itself (or any other function) a new *frame* (chunk of space) is allocated on the stack and the new parameters and the return address are saved there. This way there is no confusion what to work with and where to return when the function has finished.

> **For a recursive function to work, it must have a simple *base case* for which recursive calls are not needed.**

In the above code, `n = 1` is the base case.

> **When a function is called recursively, it must be called for smaller and smaller tasks, which eventually converge to the base case (or one of the base cases).**

Each time we call the function `add_numbers` recursively, we call it with an argument that is smaller by 1 than the previous time.

Recursion has its cost: first, you need to have enough space on the stack and, second, the code spends extra time pushing and popping data and calling the same function. Iterations are often more efficient. But some applications, especially those dealing with nested structures or branching processes, require a stack anyway, and recursion allows you to write very concise and clear code.

Example 1

Write a recursive function that calculates 10^n for a non-negative integer n.

Solution

```python
def pow10(n):
    """Return 10^n."""
    if n == 0:  # base case
        return 1
    else:
        return pow10(n-1) * 10
```

Example 2

Write a recursive function that returns the sum of the digits in a non-negative integer.

Solution

```python
def sum_digits(n):
    """Return the sum of the digits in n."""
    if n < 10: # Base case
        return n
    n, d = divmod(n, 10)   # n is now n//10
    return sum_digits(n) + d
```

Example 3

Write a recursive function that takes a string and returns a reversed string.

Solution

```python
def reversed(s):
    """Return s reversed."""
    if len(s) > 1:
        s = reversed(s[1:]) + s[0]
    return s
```

The base case here is implicit: when the string is empty or consists of only one character, there is nothing to do.

❖   ❖   ❖

A very popular example of recursion in computer programming books and tutorials is the "Tower of Hanoi" puzzle. We have three pegs and *n* disks of increasing size. Initially all the disks are stacked in a tower on one of the pegs (Figure 16-1). The objective is to move the tower to another peg, one disk at a time, without ever placing a bigger disk on top of a smaller one.

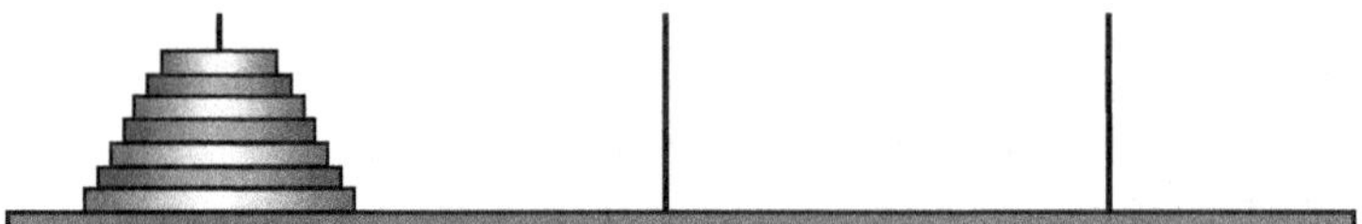

Figure 16-1. The Tower of Hanoi puzzle

The key to the solution is to notice that in order to move the whole tower, we first need to move a smaller tower of $n-1$ disks to the spare peg, then move the bottom disk to the destination peg, then move the tower of $n-1$ disks from the spare peg to the destination peg (Figure 16-2).

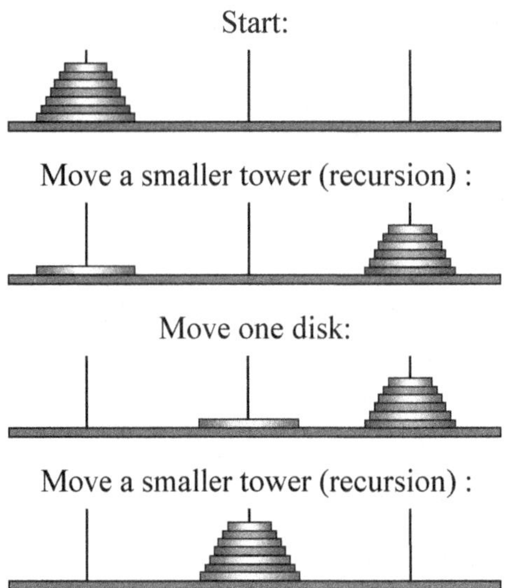

Figure 16-2. A recursive solution to Tower of Hanoi

Example 4

Write a recursive function that solves the Tower of Hanoi puzzle for n disks.

Solution

```python
def move_tower(n, from_peg, to_peg):
    """Print the solution to Tower of Hanoi."""
    # (This code just prints out each move.)
    spare_peg = 6 - from_peg - to_peg  # See Question 5
    if n > 1:
        move_tower(n - 1, from_peg, spare_peg)
    print('From', from_peg, 'to', to_peg)
    if n > 1:
        move_tower(n - 1, spare_peg, to_peg)
```

If we want to move a tower of, say, seven disks from peg 1 to peg 2, an initial call would be `move_tower(7, 1, 2)`.

Note that in the above example the function does not return a value (more precisely, it returns `None`), so it is a *recursive procedure*. Programming this task without recursion would be more difficult.

Section 16.3 ~ Exercises

1. Write and test a recursive function that returns *n*! (n-factorial). Do not use any loops. ⸮ Hint: Recall that 0! is defined to be 1. ⸝ ✓

2. Pretend that Python's `bin` function does not exist and write a <u>recursive</u> function `to_bin(n)` (without any loops) that takes a positive integer and returns a string of its binary digits. For example, `to_bin(5)` should return `'101'`. ⸮ Hint: The base cases are n = 0 and n = 1. ⸝

3. Write and test a recursive version of `eval_polynomial(p, x)`, based on the formula $a_n x^n + a_{n-1} x^{n-1} + ... + a_1 x + a_0 = x\left(a_n x^{n-1} + a_{n-1} x^{n-1} + ... + a_1\right) + a_0$. ✓

4. (a) Write and test a function `print_digits(n)` that displays a triangle with *n* rows, made of digits. For example, `print_digits(5)` should display

```
55555
4444
333
22
1
```

The function's code under the docstring should be three lines. ✓

(b) The same as Question 4 (a), but invert the triangle, so that
`print_digits(5)` prints

```
1
22
333
4444
55555
```

5.   Explain the statement `spare_peg = 6 - from_peg - to_peg` in Example 4. ✓

6.   Identify the base case in the `move_tower` function in Example 4. ✓

7.   Without running the code in Example 4, determine the output when `move_tower(3, 1, 2)` is called.

8.   The Tower of Hanoi puzzle was invented by the French mathematician Édouard Lucas in 1883. The "legend" that accompanied the puzzle stated that in Benares, India, there was a temple with a dome that marked the center of the world. The Hindu priests in the temple moved golden disks between three diamond needles. God placed 64 gold disks on one needle at the time of Creation, and the universe will come to an end when the priests have moved all 64 disks to another needle. What is the total number of moves needed to move a tower of 2 disks? 3 disks? 4 disks? n disks? Assuming one move per second, estimate the "lifespan of the universe." ✓

9.   (a) Write your own recursive function that efficiently calculates x^n, using the following property: $x^n = \left(x^k\right)^2$ if $n = 2k$ and $x^n = \left(x^k\right)^2 \cdot x$ if $n = 2k+1$.

(b)■ Write an iterative version of the same function using the binary representation of n. ⸮ Hint: Suppose $n = b_0 + b_1 \cdot 2 + b_2 \cdot 2^2 + \ldots + b_k \cdot 2^k$, where $b_0, b_1, b_2, \ldots, b_k$ are the binary digits of n. Then $x^n = x^{b_0} \cdot \left(x^2\right)^{b_1} \cdot \left(x^4\right)^{b_2} \ldots \cdot \left(x^{2^k}\right)^{b_k}$. Set the result to 1, then if bit k in n is set (counting the bits from $k = 0$ for the rightmost bit) multiply the result by the factor $f = x^{\left(2^k\right)}$. So f is initially set to x, then squared on each iteration. ⸮

10.♦♦ Find the last 9 digits of $F_{10^{12}}$, the Fibonacci number with the index 10^{12} , assuming $F_1 = 1$, $F_2 = 1$.

⸻ Hints:

1. A sequence defined by a linear recurrence relation, such as Fibonacci numbers, can be computed using matrices. Recall that when a matrix

$$A = \begin{pmatrix} a_{11} & a_{12} \\ a_{21} & a_{22} \end{pmatrix}$$ is applied to a vector $v = \begin{pmatrix} x \\ y \end{pmatrix}$, the result is a new vector,

$$Av = \begin{pmatrix} a_{11}x + a_{12}y \\ a_{21}x + a_{22}y \end{pmatrix}.$$ If $A = \begin{pmatrix} 0 & 1 \\ 1 & 1 \end{pmatrix}$, then $A \begin{pmatrix} F_{n-1} \\ F_n \end{pmatrix} = \begin{pmatrix} F_n \\ F_{n+1} \end{pmatrix}$. If you apply A again, you get the next pair of Fibonacci numbers. If you apply A $(n-1)$ times to $\begin{pmatrix} F_1 \\ F_2 \end{pmatrix}$, you get $\begin{pmatrix} F_n \\ F_{n+1} \end{pmatrix}$. In other words,

$$\begin{pmatrix} F_n \\ F_{n+1} \end{pmatrix} = A^{n-1} \begin{pmatrix} F_1 \\ F_2 \end{pmatrix}.$$

2. The efficient x^n function from Question 9 can be adapted for raising a matrix to the n-th power and will work very fast for a 2 by 2 matrix and $n = 10^{12}$. When multiplying matrices, keep track only of the last 9 digits.

16.4 Mathematical Induction

Let d_n be the number of moves required to move a tower of n disks in the Tower of Hanoi puzzle. It is pretty obvious that, for $n > 1$, $d_n = d_{n-1} + 1 + d_{n-1} = 2d_{n-1} + 1$, because to move a tower of n disks, we need to first move a tower of $n-1$ disks, then one disk, then again a tower of $n-1$ disks. If you finished Question 8 in the previous section, you probably guessed correctly that $d_n = 2^n - 1$. How can we prove this fact more rigorously? In questions of this kind we can use the method of proof called *mathematical induction*.

Suppose you have two sequences, $\{a_n\}$ and $\{b_n\}$. Suppose that:

1. $a_1 = b_1$ (base case).

2. For <u>any</u> $n > 1$ you have managed to establish the following fact: <u>if</u> $a_{n-1} = b_{n-1}$ <u>then</u> $a_n = b_n$.

Is it then true that $a_n = b_n$ for all $n \geq 1$?

Let's see: we know that $a_1 = b_1$. We know that if $a_1 = b_1$ then $a_2 = b_2$. So we must have $a_2 = b_2$. Next step: we know that $a_2 = b_2$. We know that if $a_2 = b_2$ then $a_3 = b_3$. So we must have $a_3 = b_3$. Next step: ... and so on. We have to conclude that $a_n = b_n$ for all $n \geq 1$. In the future, we do not have to repeat this chain of reasoning steps every time — we just say our conclusion is true "by mathematical induction."

Example 1

Prove that the number of moves in the Tower of Hanoi puzzle for n disks $d_n = 2^n - 1$.

Solution

d_n satisfies the relationship

$$d_1 = 1,$$
$$d_n = 2d_{n-1} + 1 \text{ for } n > 1$$

We want to show that $d_n = 2^n - 1$ for all $n \geq 1$.

1. First let's establish the base case, $n = 1$. $d_1 = 1$ and $2^1 - 1 = 1$, so $d_1 = 2^1 - 1$. OK.

2. Now let's proceed with the *induction step.* <u>Assume</u> that $d_{n-1} = 2^{n-1} - 1$. Then $d_n = 2d_{n-1} + 1 = 2(2^{n-1} - 1) + 1 = 2^n - 2 + 1 = 2^n - 1$. So, for any $n > 1$, we were able to show that <u>if</u> $d_{n-1} = 2^{n-1} - 1$ <u>then</u> $d_n = 2^n - 1$.

By mathematical induction, $d_n = 2^n - 1$ for all $n \geq 1$, Q.E.D.

The method of mathematical induction applies to any sequence of propositions $\{P_n\}$. Suppose we know that P_1 is true; suppose we can show for any $n > 1$ that <u>if</u> P_{n-1} is true <u>then</u> P_n is true. Then, by mathematical induction, P_n is true for all $n \geq 1$. (In the above example, P_n is the proposition $d_n = 2^n - 1$.)

Sometimes it is necessary to use a more general version of mathematical induction, with several base cases and a more general induction step. Suppose that:

1. $P_1, ..., P_k$ are true (base case(s)).

2. For any $n > k$ we can prove that <u>if</u> $P_1, P_2, ..., P_{n-1}$ are all true, <u>then</u> P_n is true.

Then, by mathematical induction, P_n is true for all $n \geq 1$.

Example 2

Prove that the *n*-th Fibonacci number $F_n \geq \left(\dfrac{3}{2}\right)^{n-2}$.

Solution

1. We can establish two base cases:

$$F_1 = 1 \geq \frac{2}{3} = \left(\frac{3}{2}\right)^{-1} = \left(\frac{3}{2}\right)^{1-2}$$

$$F_2 = 1 \geq \left(\frac{3}{2}\right)^{0} = \left(\frac{3}{2}\right)^{2-2}.$$

continued ➷

2. Now the induction step. Assume that for any $m < n$ $F_m \geq \left(\dfrac{3}{2}\right)^{m-2}$. In

particular, for $n > 2$, $F_{n-1} \geq \left(\dfrac{3}{2}\right)^{n-3}$ and $F_{n-2} \geq \left(\dfrac{3}{2}\right)^{n-4}$. Then

$$F_n = F_{n-1} + F_{n-2} \geq \left(\dfrac{3}{2}\right)^{n-3} + \left(\dfrac{3}{2}\right)^{n-4} = \left(\dfrac{3}{2}\right)^{n-4}\left(\dfrac{3}{2}+1\right) =$$

$$\left(\dfrac{3}{2}\right)^{n-4} \cdot \dfrac{5}{2} > \left(\dfrac{3}{2}\right)^{n-4} \cdot \dfrac{9}{4} = \left(\dfrac{3}{2}\right)^{n-2}$$

So the proposition is true for the two base cases, and <u>if</u> it is true for $n-1$ and $n-2$ <u>then</u> it is also true for n. By math induction, it is true for all $n \geq 1$, Q.E.D.

The 100-th Fibonacci number is $F_{100} = 354224848179261915075$. The above result explains why Fibonacci numbers grow so fast.

Section 16.4 ~ Exercises

1.   Prove using mathematical induction that $1 + 3 + 5 + ... + (2n - 1) = n^2$. ✓

2.■   Consider the following function:

```python
def tangle(s):
    n = len(s)
    if n < 2:
        return ''   # empty string
    else:
        return tangle(s[1:n]) + s[1:n-1] + tangle(s[0:n-1])
```

Show that `tangle('abcde')` returns a string that contains neither `'a'` nor `'e'`.

3.◆   Show that for a string s of length n, `tangle(s)`, defined in Question 2, returns a string of length $L_n = 2^{n-1} - n$. ⸮ Hint: first obtain a recurrence relation for L_n, then prove it using mathematical induction. ⸮ ✓

4.◆   Suppose you have n straight lines in a plane, such that no two lines are parallel to each other and no three lines go through the same point. Show that the number of regions into which these lines cut the plane depends only on n, but not on a particular configuration of the lines. Find a formula for that number and prove it using mathematical induction. ⸮ Hint: when you add a line, the existing lines cut it into pieces. Each of the pieces of the new line cuts a region into two, adding a new region. ⸮ ✓

5.◆   Consider the sequence $a_0 = 2$, $a_1 = 2$, $a_n = 2a_{n-1} + 3a_{n-2}$ for $n \geq 2$. Write a Python program that prints out the first n terms of this sequence. Examine the pattern and come up with a closed-form formula for a_n, then prove that your guess is correct using mathematical induction. ⸮ Hint: compare a_n to 3^n. ⸮

6.◆   Consider a recursive version of the function that returns the n-th Fibonacci number:

```python
def fibonacci(n):
    """Return the n-th Fibonacci number."""
    if n == 1 or n == 2:
        return 1
    return fibonacci(n-1) + fibonacci(n-2)
```

Test it with $n = 6$ and $n = 20$. Now test it with $n = 100$. The computer won't return the result right away. Perhaps the recursive version takes a little longer... Prove by mathematical induction that the total number of calls to `fibonacci`, including the original call and all the recursive calls combined, is not less than the n-th Fibonacci number F_n. We have shown above that

$$F_n \geq \left(\frac{3}{2}\right)^{n-2}$$. Assuming that your computer can execute one trillion calls per second, estimate how long you will have to wait for `fibonacci(100)` (in years). Press Ctrl-C to abort the program. As we said earlier, sometimes recursion can be costly.

16.5 Review

Terms introduced in this chapter:

> *Recurrence relation*
> *Recursion*
> *Stack*
> *Recursive function*
> *Recursive procedure*
> *Base case*
> *Mathematical induction*

Python feature introduced in this chapter:

> Recursive functions
> ```
> list.append(x)
> list.pop()
> ```

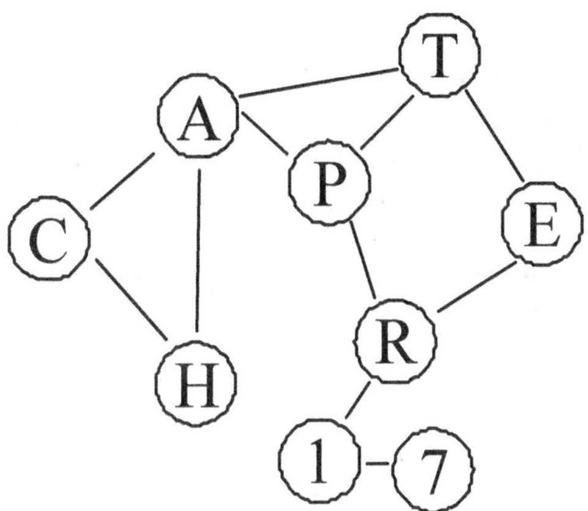

Graphs

17.1 Prologue

In the 1700s, the city of Königsberg, then the capital of East Prussia near the Baltic Sea, had seven bridges puzzle connecting the banks of the river Pregel with the banks of its tributary and the Kniephof island in the center. No one remembers for sure how the following puzzle came about: to find a path through the city that crosses each of the seven bridges exactly once (Figure 17-1). Legend has it that many citizens took long walks around the city on Sundays trying to solve the puzzle. (Believe it or not, there were many lovers of puzzles in the 1700s, just as there are today.)

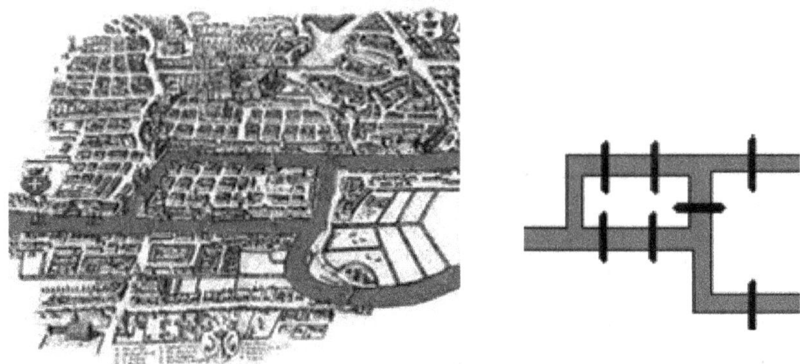

Figure 17-1. The Seven Bridges of Königsberg puzzle

Can you find a solution?

Finally, around 1735, the great Swiss mathematician Leonhard Euler (pronounced "oiler") solved the problem (as he often did when he was interested in one). In the process, Euler founded two new branches of mathematics: topology and graph theory.

To begin with, Euler simplified the picture: he compressed each piece of land into one point and replaced each bridge with a line segment that connected two points, obtaining what is now called a *graph* with four *vertices* and seven *edges* (Figure 17-2).

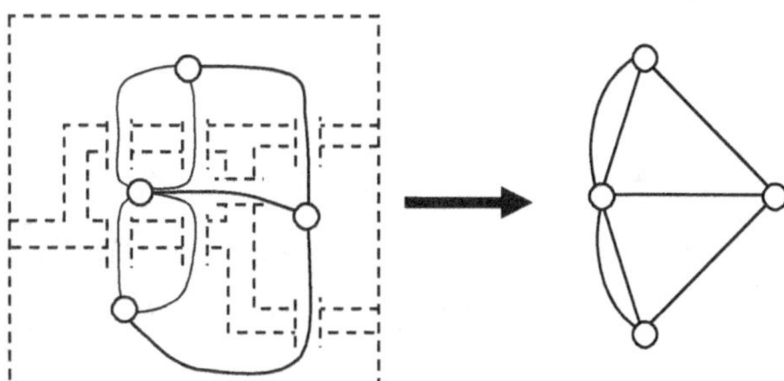

Figure 17-2. The Seven Bridges puzzle represented as a graph

The exact shape of the graph is not important — you can move the vertices around and stretch or bend the edges, as long as the *topology* of the graph (the particular pairs of vertices connected by edges) remains the same. The task is now to find a continuous path in the graph that traces each edge exactly once. Such a path is called an *Euler path*. An Euler path that starts and ends at the same vertex is called an *Euler circuit*. Before you tackle the Seven Bridges puzzle, try to find an Euler path in several other graphs (Figure 17-3) and come up with a rule to determine which graphs have an Euler path or an Euler circuit and which don't (see Question 9 in Section 17.4).

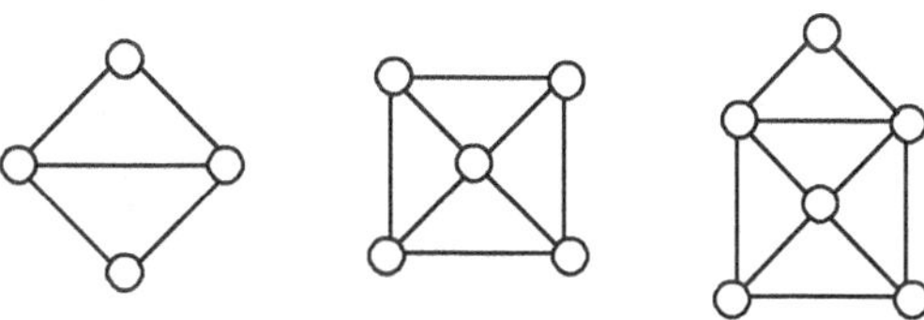

Figure 17-3. Examples of graphs

A graph can be used to represent a computer network, a system of routes for an airline, a road map, a set of possible positions and legal moves in a game, and so on (Figure 17-4). Thus graphs serve as a universal modeling tool.

Graph theory is a vast and fascinating branch of mathematics; here we can only give you a first taste of it. The basic definitions are simple, making it a great playground for amateur mathematicians of all ages. But some of the theorems are very hard.

In the last two sections we will talk about coloring geographic maps and planar graphs, which, as we will see, is the same thing, and then try to tackle the Four Color Theorem for planar graphs.

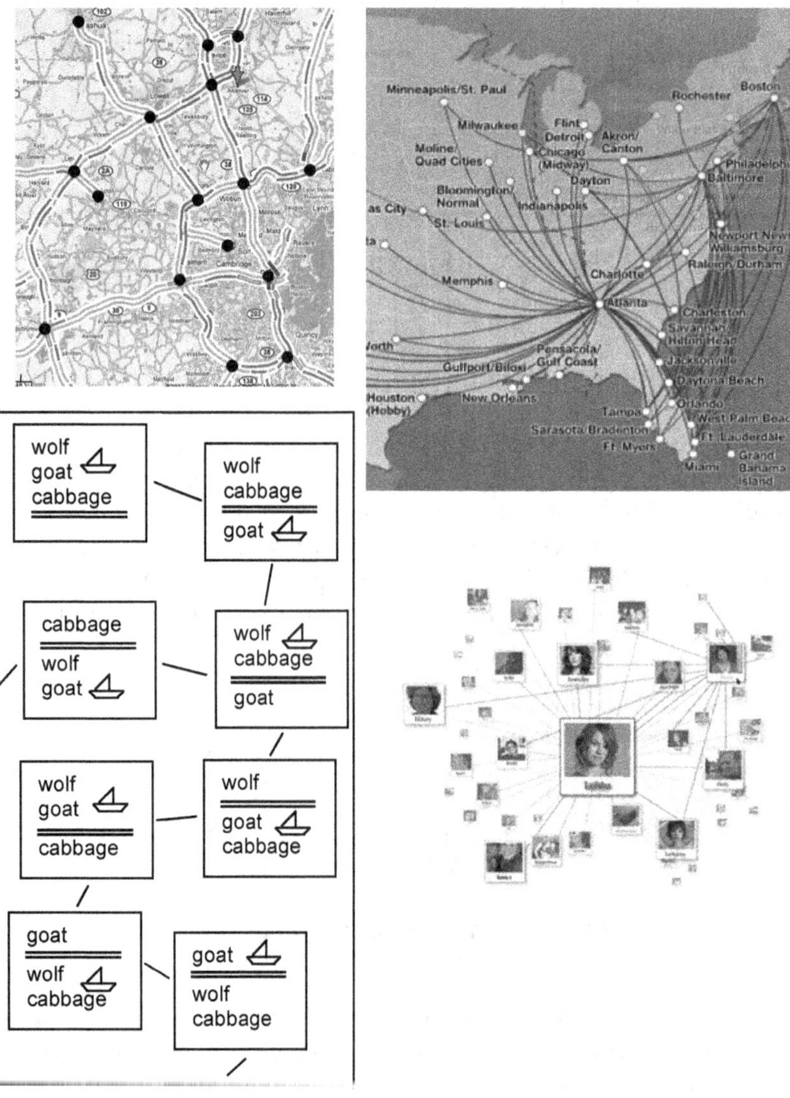

Figure 17-4. Some applications of graphs

17.2 Types of Graphs

A graph can be described in an abstract way, as a set of its *vertices* and a set of its *edges*. An edge is represented by an unordered pair of vertices; it is understood that the edge "connects" that pair of vertices. The vertices of a graph are sometimes called *nodes*, and the edges are sometimes called *arcs*.

Figure 17-3 shows graphs as points on a plane connected by line segments. Sometimes mathematicians indeed work with such *planar graphs*. But in general, a drawing of a graph simply helps to visualize it.

A graph can have several edges connecting the same pair of vertices. Such graphs are called *multigraphs*. The Seven Bridges graph (Figure 17-2) is an example. A graph can also have an edge connecting a vertex to itself (Figure 17-5). Such edges are called *loops*. In a *simple graph* (such as those shown in Figure 17-3), no more than one edge connects a pair of vertices, and there are no loops.

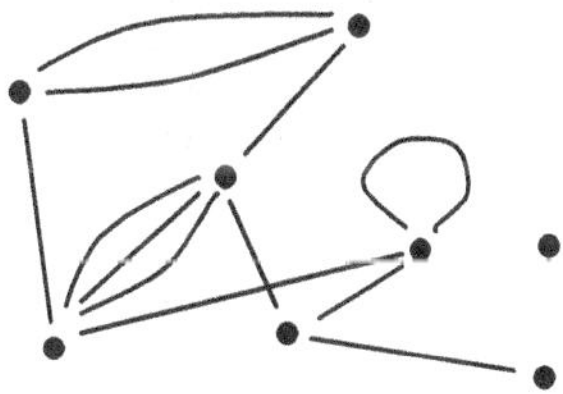

**Figure 17-5. A *multigraph*: several edges can connect the
same pair of vertices, and some vertices can be
connected to themselves**

Example 1

Describe the simple graph below as a set of vertices and a set of edges.

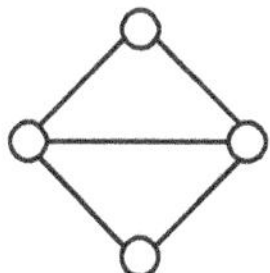

Solution

First we need to label the vertices of the graph to be able to refer to them. Let's use the numbers from 1 to 4:

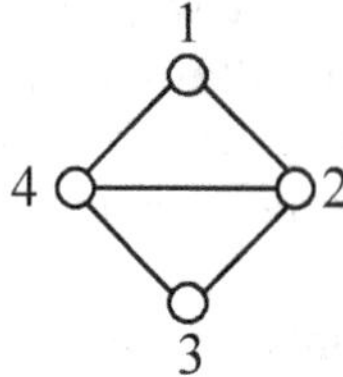

Now $V = \{1,2,3,4\}$, $E = \{(1,2), (2,3), (3,4), (4,1), (2,4)\}$, $G = (V,E)$. Or, in Python:

```
v = {1, 2, 3, 4}   # a set
e = {(1, 2), (2, 3), (3, 4), (4, 1), (2, 4)}
g = (v, e)
```

❖   ❖   ❖

When a graph is used as a model for a real-life situation, its vertices may represent different geographical locations, different objects, or different positions in a game. But if you view a graph as an abstract mathematical object, the graph is defined only by its *topology*; that is, by the configuration of its vertices and edges.

> **A simple graph in which every pair of vertices is connected by an edge is called a *complete graph*.**

A complete graph with *n* vertices is usually denoted by K_n.

Example 2

Draw K_4 and K_5.

Solution

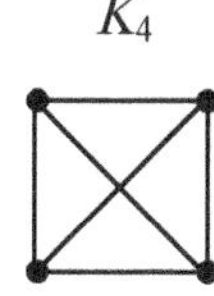
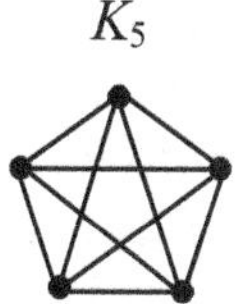

A complete graph could be drawn to look different, but from the point of view of graph theory it would be essentially the same graph as long as it had the same number of vertices, and exactly one edge connected every vertex to every other vertex.

Another simple type of graph is one in which the vertices are arranged into a circular sequence and each vertex is connected to two of its neighbors. (Think of a circle of people holding hands.) Such a graph is called a *cycle*. A cycle with n vertices is usually denoted by C_n.

Example 3

Draw C_6.

Solution

Section 17.2 ~ Exercises

1.   Find an Euler path in the graph below. ✓

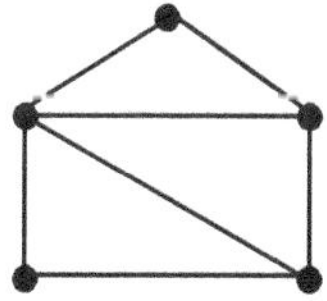

2. Draw a sketch of the graph $G = (V, E)$, where $V = \{a, b, c, d\}$ and
$E = \{(a, b), (a, c), (b, c), (b, d)\}$. ✓

3. Label the vertices of the graph below with letters or numbers, then describe the graph as a set of its vertices and a set of its edges.

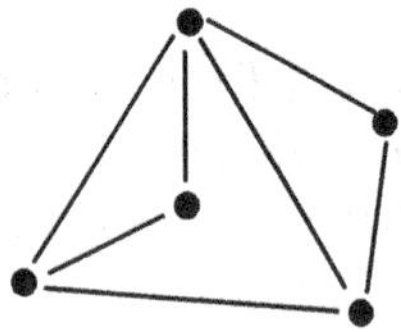

4. Which of the following graphs are simple graphs? Which are multigraphs? Which have loops? ✓

(a)   (b)   (c)   (d)

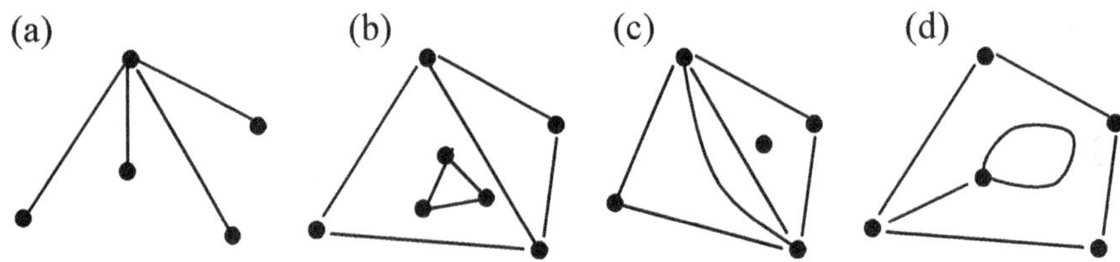

5. (a) How many edges does C_{12} have?

(b) How many edges does K_{12} have? ✓

(c) For which n is C_n the same as K_n?

17.3 Isomorphism of Graphs

Two graphs are called *isomorphic* if there exists a one-to-one correspondence between their vertices and a one-to-one correspondence between their edges, such that two vertices are connected by an edge in the first graph if and only if the corresponding vertices are connected by the corresponding edge in the second graph.

Example 1

Consider two graphs: $G_1 = (V_1, E_1)$ and $G_2 = (V_2, E_2)$. The set of the vertices of the first graph is $V_1 = \{A, B, C, D, E\}$; the set of its edges is $E_1 = \{(A,B), (A,C), (B,C), (B,D), (C,E), (D,E)\}$. For the second graph, $V_2 = \{1, 2, 3, 4, 5\}$ and $E_2 = \{(1,2), (2,3), (3,4), (4,5), (5,1), (2,4)\}$. Are G_1 and G_2 isomorphic?

Solution

To find out, let's draw the two graphs:

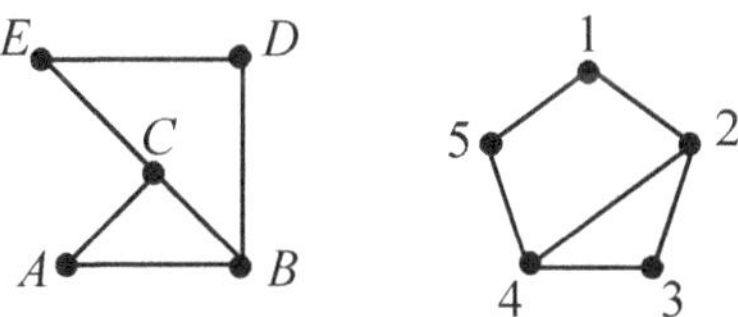

At a first glance, they seem different, but a little bending (without breaking the edges) will convince us that they in fact have the same configuration:

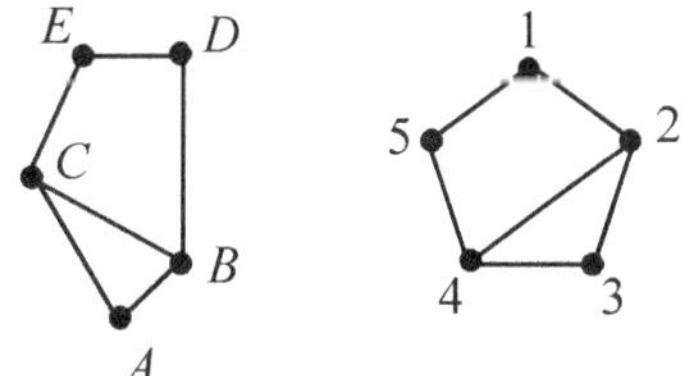

The correspondence $A \leftrightarrow 3$, $B \leftrightarrow 2$, $C \leftrightarrow 4$, $D \leftrightarrow 1$, $E \leftrightarrow 5$ establishes an isomorphism between these two graphs.

Note how we labeled the vertices of a graph to be able to refer to them. We used letters for the first graph and numbers for the second. In general we can use any labels or names.

Isomorphism between graphs has three properties:

1. *Reflexivity*: any graph is isomorphic to itself;
2. *Symmetry*: if G_1 is isomorphic to G_2, then G_2 is isomorphic to G_1;
3. *Transitivity*: if G_1 is isomorphic to G_2 and G_2 is isomorphic to G_3, then G_1 is isomorphic to G_3.

> **A relationship that has these three properties is called an *equivalence relation*.**

An equivalence relation on the elements of a set breaks that set into non-overlapping subsets, called *equivalence classes*. Any two elements in the same equivalence class are "equivalent" to each other (that is, the equivalence relation holds for them), and any two elements in different equivalence classes are not "equivalent."

Since isomorphism is an equivalence relation between graphs, all graphs fall into non-overlapping classes of isomorphic graphs.

Another example of an equivalence relation is the "being connected by a path" relation between two vertices of a graph G. This relation is true for the vertices X and Y if there is a path from X to Y. (An empty "path" connects a vertex to itself.) It is easy to see that the three properties of an equivalence relation are satisfied for this relation. In particular, if X is connected by a path to Y, and Y is connected by a path to Z, then X is connected by a path to Z (just combine the paths — Figure 17-6).

> **A graph is said to be *connected* if, for any two of its vertices, there is a path connecting them.**

An equivalence class for the "being connected by a path" relation is a set of vertices in a connected subgraph of G. Therefore, any graph is a union of non-intersecting connected subgraphs (Figure 17-6).

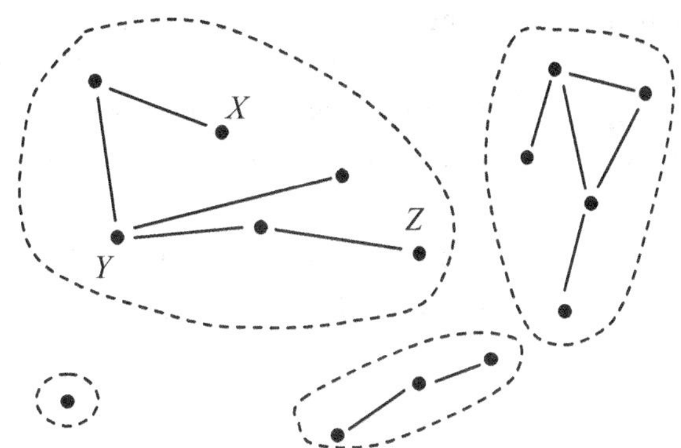

Figure 17-6. A graph can be split in a unique way into non-intersecting connected subgraphs

Section 17.3 ~ Exercises

1.   Determine whether the following pairs of graphs are isomorphic: ✓

(a)

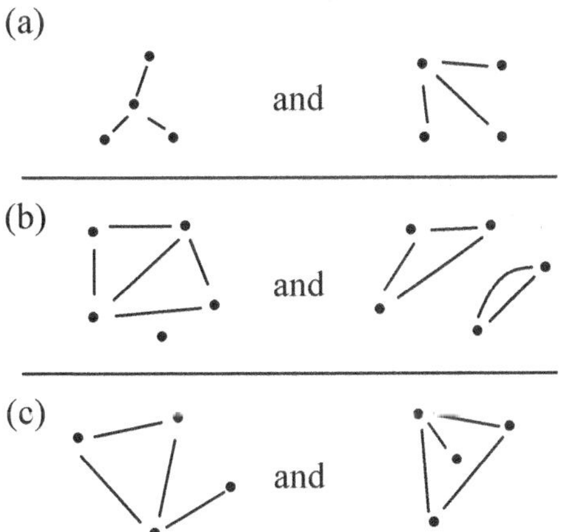

(b)

(c)

2.   Is there a different correspondence of vertices for the graphs in Example 1 that demonstrates their isomorphism? If yes, show it; if not, explain why not. ✓

3.   How many different simple graphs with four vertices are there, if you count all isomorphic graphs as one? How many of them are connected? ✓

4.   A graph $G_2 = (V_2, E_2)$ is called a *subgraph* of $G_1 = (V_1, E_1)$ if $V_2 \subseteq V_1$ and $E_2 \subseteq E_1$. Can a graph be isomorphic to its subgraph? Explain. ✓

5.■   Does K_n have a subgraph isomorphic to C_n? Does K_n have a subgraph isomorphic to C_{n-2} when $n \geq 5$? Does K_n have a subgraph isomorphic to K_{n-2} when $n \geq 5$?

6.■   Is $x \leq y$ an equivalence relation on the set of all real numbers?

7.■   Come up with five examples of equivalence relations between two integers.

8.■   Come up with an example of an equivalence relation between two sets (other than equality). ✓

17.4   Degree of a Vertex

The number of edges that come out of a vertex of a graph is called the *degree* of that vertex.

If an edge is a loop (connects a vertex to itself), that edge is counted twice in the degree of that vertex. In a simple graph, the degree of a vertex is simply the number of other vertices to which this vertex is connected.

Example 1

What are the degrees of the vertices in the following multigraph?

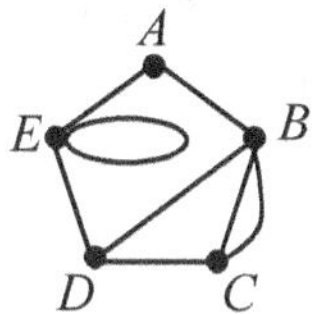

Solution

A: 2; B: 4; C: 3; D: 3; E: 4.

If two graphs are isomorphic, the degrees of the corresponding vertices must match, of course.

Section 17.4 ~ Exercises

1. If a graph has n vertices, and $d_1, d_2, ..., d_n$ are their degrees, what is the total number of edges in that graph? ✓

2. Is it possible to have a graph with five vertices whose degrees are 2, 2, 4, 3, and 6? If yes, give an example; if not, explain why not. ✓

3. Can a graph have five vertices, each connected to exactly three others? Draw an example or explain why not.

4. Is it possible to have a group of five colleagues, each corresponding via e-mail with exactly three others?

5. Prove that the only simple graph with n vertices and the degrees of all the vertices equal to $n-1$ is K_n. ✓

6.■ Prove that the only simple connected graph with three or more vertices and the degrees of all the vertices equal to 2 is C_n. ✓

7. Give an example of two graphs with the same number of vertices, such that the vertices in the first graph can be ordered so that their degrees match the degrees of the vertices in the second graph, but the graphs are not isomorphic to each other.

8. Show that in any group of six people there are either three who all know each other or three who are all strangers to each other.   Hint:

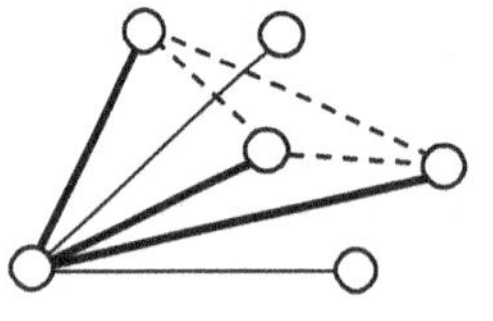

9. Recall that an Euler path traverses all the edges of a graph and passes through each edge exactly once. Show that if a graph has an odd-degree vertex, then an Euler path, if it exists, must begin or end at that vertex.

10. Remove one bridge in the Seven Bridges of Königsberg puzzle so that it becomes solvable. ✓

11. We want to make a wire framework for a cube. We want to bend a single piece of wire so that it covers as many edges of the cube as possible, without doubling over any edge. What is the smallest number of edges that will remain missing?

12. In a graph, suppose L is the number of vertices whose degrees are odd. Describe, in terms of L, a necessary and sufficient condition for the graph to have an Euler path. ✓

13. An *Euler circuit* in a graph begins and ends at the same vertex and traverses every edge of the graph exactly once. Describe the necessary and sufficient condition for a graph to have an Euler circuit.

14. Show that for any $n \geq 3$ there is a connected graph with n vertices that has an Euler circuit (as defined in Question 13). ✓

15. For which $n \geq 3$ does K_n have an Euler circuit?

16. Give an example of a connected graph that cannot be turned into a graph with an Euler circuit by removing one or several edges. ✓

17. A *Hamilton circuit* in a graph starts and ends at the same vertex, goes along the edges of the graph, and visits each vertex exactly once. Hamilton circuits are named after an Irish mathematician, Sir William Rowan Hamilton (1805-1865). In 1859, Hamilton designed a puzzle and sold it to a toy maker in Dublin. The puzzle was a regular *dodecahedron* (a solid with twelve pentagon-shaped faces and 20 vertices), made of wood. Each vertex was labeled with the name of a prominent city. A player had to find a path beginning and ending at the same vertex that went along the edges and visited each "city" exactly once. Solve Hamilton's original puzzle. ⸙ Hint: you can work with pencil and paper using a planar graph isomorphic to the dodecahedron:

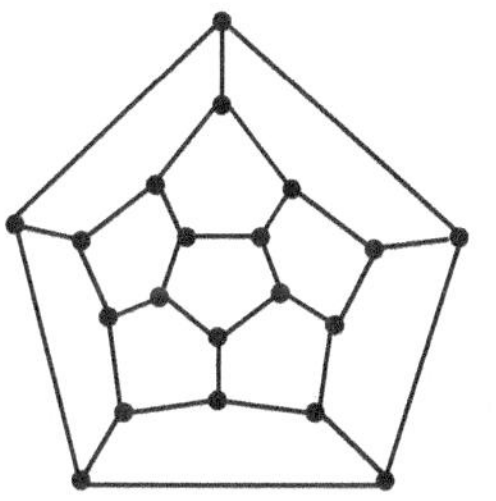

18. Can you find a Hamilton circuit (as defined in Question 17) in the following graph?

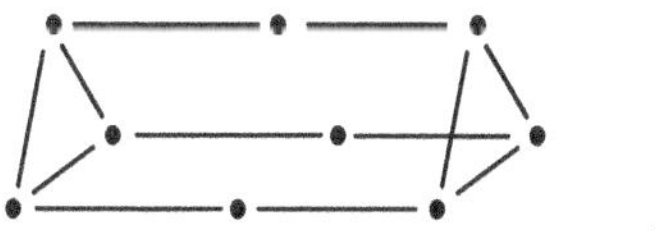

19. Write a function `has_euler_path(g)` that takes a graph `g`, represented as a pair `(V, E)`, where `V` is the set of vertices and `E` is the set of edges (pairs of vertices) and returns `True` if `g` has an Euler path; otherwise your function should return `False`.

17.5 Directed and Weighted Graphs

The graphs described in the previous sections are called *undirected*, because their edges have no direction: a vertex A is connected to a vertex B if and only if B is connected to A. We can also consider *directed graphs*, in which edges are <u>arrows</u> that show the direction of the connection (Figure 17-7).

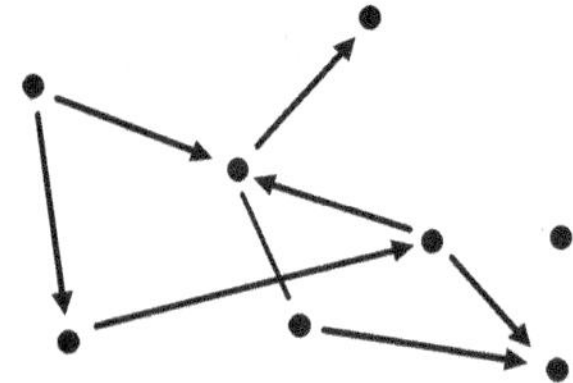

Figure 17-7. A directed graph

In a directed graph, there may be an arrow from vertex A to vertex B but not from B to A, or there may be arrows in both directions (Figure 17-8).

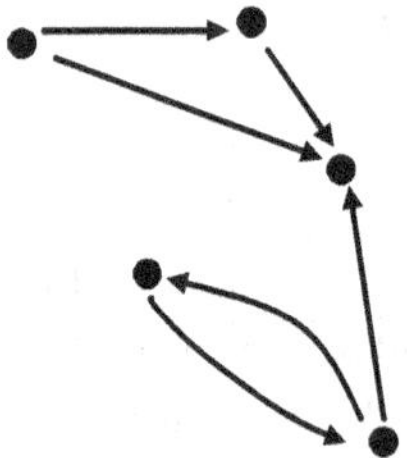

**Figure 17-8. A directed graph with a pair of vertices
connected in both directions**

If we have a directed graph, we can associate with it an undirected graph, replacing each arrow with an edge.

Example 1

Draw an undirected graph associated with the directed graph in Figure 17-8.

Solution

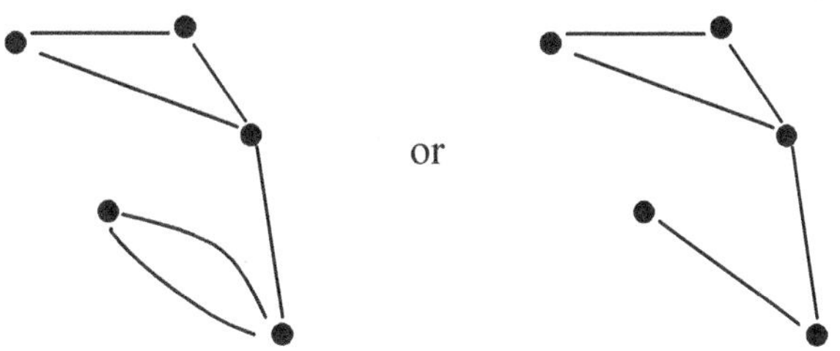

or

depending on how we agree to handle multiple edges between the same two vertices.

> **A directed graph is called *connected* if the associated undirected graph is connected.**

Some typical tasks involving directed graphs are to find whether a path exists from one vertex to another along the arrows or to find the shortest such path.

A *weighted graph* is a graph with numbers ("weights") associated with its edges. For example, a transit network can be represented by a weighted graph in which the weights are distances between adjacent stations. A computer network can be represented by a weighted graph in which weights represent the costs of transmitting information over the links.

A typical task for a weighted graph is to find an *optimal path* between two given vertices. The sum of the weights along such a path must be the smallest possible.

Example 2

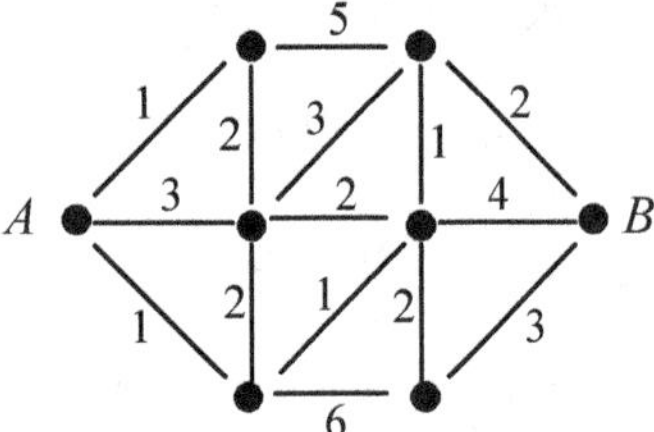

For the weighted graph above, find an optimal path from *A* to *B*.

Solution

You can try to list all possible paths, of course, but there are too many of them. A more promising approach is to first find optimal paths to *B* from each <u>neighbor</u> of *A* and mark the cost of each such optimal path:

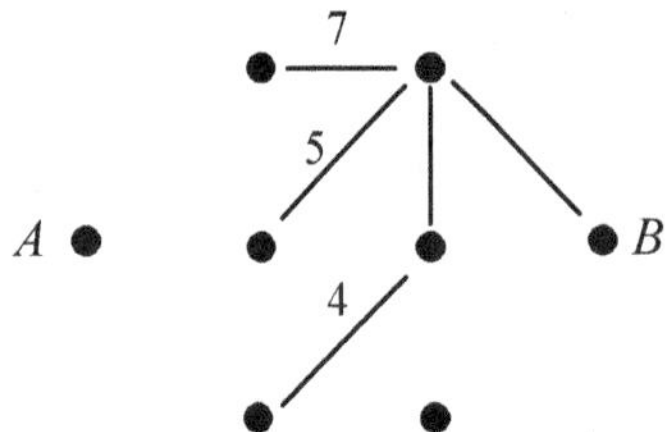

Now we can add the cost of going from *A* to a neighbor to the cost of going from that neighbor to *B* and find the best combination:

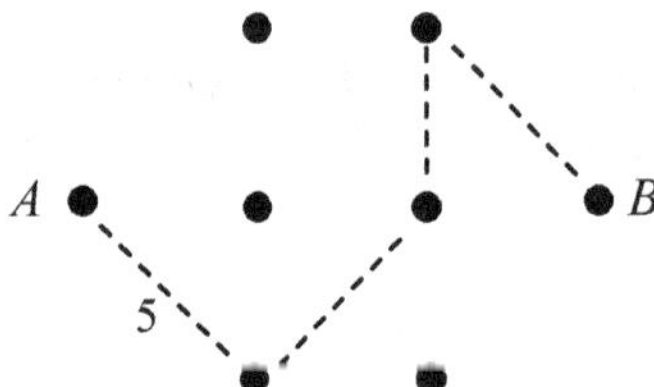

We can do this manually for a small graph, but obviously we'll need a computer program for larger graphs.

Section 17.5 ~ Exercises

1.   Define isomorphism for directed graphs. ✓

2.   A cycle in a directed graph is a path along the arrows that starts and ends at the same vertex. Show that if every vertex in a directed graph has at least one arrow coming out of it, then the graph has a cycle.

3.■   Find an optimal path (with the smallest sum of weights along it) from A to B in the following directed graph: ✓

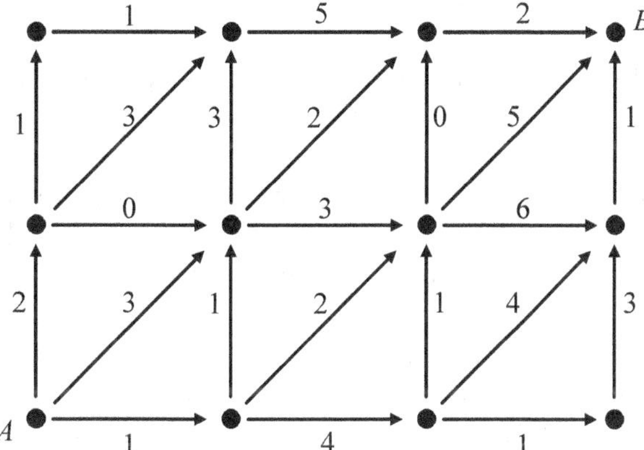

4.■   Four people need to cross a bridge in the dark. They all begin on the same side and they have only one flashlight. A maximum of two people can cross at one time. Any party that crosses, either 1 or 2 people, must have the flashlight with them. The flashlight must be walked back and forth (it may not be thrown, etc.). The four people walk at different speeds: it takes them 1, 2, 5, and 10 minutes, respectively, to cross from one side to the other. What is the optimal strategy and the minimum time required for bringing all four people across the bridge? Can it be accomplished in less than 19 minutes? ⋦ Hint: Represent the problem as a weighted graph where the vertices represent all possible positions and the edges represent allowed crossings; assign a weight — the duration of the crossing — to each edge; find the optimal path from the initial to the final position. ⋧

5. In a famous puzzle dating back at least a thousand years, a farmer has a wolf, a goat, and a cabbage. He needs to cross a river and also transport his possessions across, but his tiny boat can hold only him and one other object or animal. How can he do it without endangering any of his possessions? The wolf would gobble up the goat in a minute, if left unattended, but she is indifferent to cabbage; the goat, in turn, would devour the cabbage. The cabbage wouldn't eat anyone.

(a) Represent the puzzle using a directed graph in which each vertex represents a certain position of the farmer and his three possessions on the two banks of the river. What is the total number of all possible positions? Leave only those vertices of the graph in which all the farmer's possessions are safe.

(b) Draw the arrows in the graph that correspond to the allowed river crossings.

(c) Find and describe the solution to the puzzle by finding a path from the initial position (all on one bank) to the final position (all on the other bank) along the edges of the graph. How many solutions are there?

6. An Euler circuit in a directed graph follows the directions of the edges. Devise a necessary and sufficient condition (in terms of the numbers of arrows coming into and going out of each vertex) for a directed graph to have an Euler circuit. ✓

17.6 Adjacency Matrices

Suppose we have a simple graph with n vertices.

> **Two vertices connected by an edge are called *adjacent*.**

One way to describe the graph's edges is simply to list all pairs of adjacent vertices. But there is another way. We can make a table with n rows and n columns and put a checkmark at the intersection of the i th row and j th column if the i th vertex is connected to the j-th vertex. In math and computer programs it is more convenient to use 0s and 1s instead of checkmarks: 1 indicates an edge and 0 no edge. A square matrix that describes the edges of a graph is called the *adjacency matrix* of that graph (Figure 17-9).

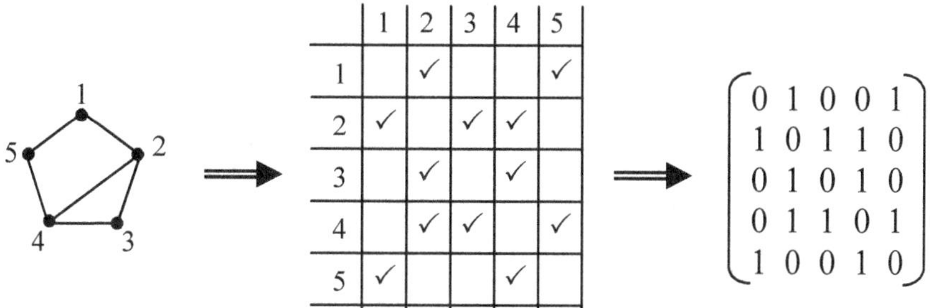

Figure 17-9. Representing a graph by its adjacency matrix

> **For a simple (not directed) graph, its adjacency matrix contains only 0s and 1s, is symmetrical over the main diagonal, and has zeros on the diagonal.**

Example 1

Write the adjacency matrix for the following graph:

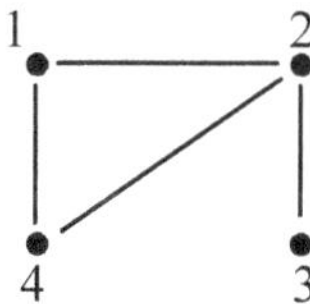

Solution

$$\begin{bmatrix} 0 & 1 & 0 & 1 \\ 1 & 0 & 1 & 1 \\ 0 & 1 & 0 & 0 \\ 1 & 1 & 0 & 0 \end{bmatrix}$$

For a directed graph, we can agree that the value a_{ij} in its adjacency matrix is 1 if there is an arrow from the i-th vertex to the j-th vertex. The adjacency matrix for a directed graph is not necessarily symmetrical.

Example 2

Write the adjacency matrix for the following directed graph:

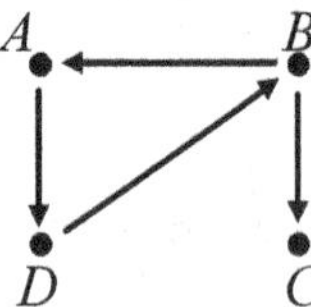

Solution

Assuming the vertices are numbered in the order A, B, C, D:

$$\begin{pmatrix} 0 & 0 & 0 & 1 \\ 1 & 0 & 1 & 0 \\ 0 & 0 & 0 & 0 \\ 0 & 1 & 0 & 0 \end{pmatrix}$$

❖   ❖   ❖

For a multigraph, an element of the adjacency matrix holds the number of edges between the corresponding vertices. For a simple weighted graph, instead of 1s and 0s we can put into the matrix the weights assigned to the edges.

Section 17.6 ~ Exercises

1.   Write the adjacency matrix for the following graph: ✓

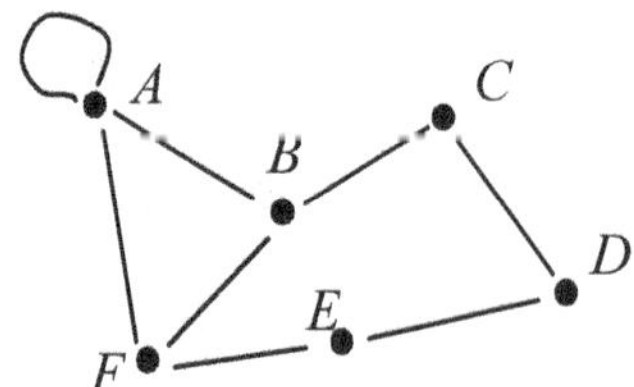

2.   Draw a directed graph that corresponds to the following adjacency matrix:

$$\begin{pmatrix} 0 & 1 & 1 & 0 \\ 0 & 0 & 1 & 0 \\ 1 & 1 & 1 & 0 \\ 0 & 1 & 0 & 0 \end{pmatrix}$$

3.   Write adjacency matrices for C_5 and K_5. ✓

4.▪   Which of the following operations on the adjacency matrix for a directed graph always results in a matrix for an isomorphic graph? (Isomorphism for graphs is defined in Section 17.3.) ✓

(a)   Flipping the matrix symmetrically over the main diagonal
(b)   Swapping any two rows
(c)   Swapping any two columns
(d)   Swapping the i-th and j-th rows, then the i-th and j-th columns (for any i and j).

5.▪ (a) Write and test a Python function that takes an adjacency matrix for a simple graph and returns a list of its edges. An edge that connects the i-th and j-th vertices should be described by the tuple `(i,j)`, where $i < j$.

(b) Modify the function from Part (a) so that it works for directed graphs. An arrow from the i-th to the j-th vertex should be described by the tuple `(i,j)`.

6.♦   Write and test a Python function that takes a simple graph $G = (V, E)$ (described by two sets, V and E) and returns its adjacency matrix.

7.♦   Suppose A is an adjacency matrix for a directed graph. How can you interpret the values of the elements of $A^2 = A \cdot A$ in terms of existing paths in the graph? ✓

8.♦   Write and test a Python function `all_paths(g, k)` that takes a directed graph `g` (described by two sets, V and E) and a positive integer k and calculates the number p_{ij} of paths of length k from the i-th vertex to the j-th vertex for all i and j. The result should be returned as a matrix with values p_{ij}. ⧼ Hint: see Questions 6 and 7. ⧽

17.7   Coloring Maps and Graphs

In a geographical map, neighboring regions, countries, or states are often shown in different colors. The map in Figure 17-10 uses five "colors" (a shade of gray for water).

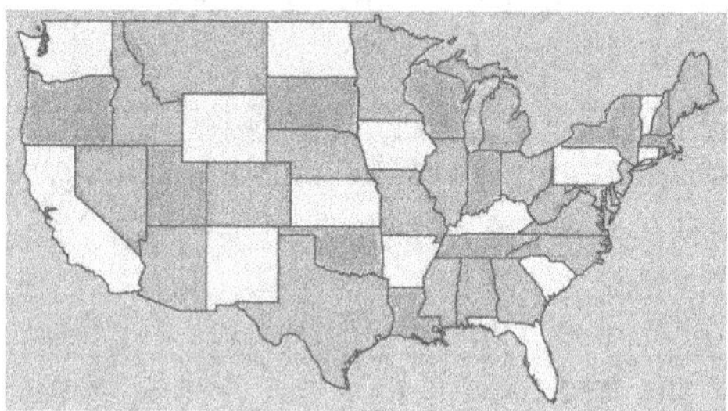

Figure 17-10.   A map colored in five "colors"

Can any map be colored in five colors? Could we use fewer colors? Four colors? Three colors? These questions have little to do with making maps: we could use as many colors as we need to make a pretty map. Instead, these questions end up in the realm of mathematics, as many interesting questions do.

Once we enter the realm of mathematics, though, we must be very precise. What is a map? What is a "country"? What does "neighboring countries" mean? Do "countries" that only touch in one point, like Colorado and Arizona at The Four Corners on a U.S. map, share a border? Can a country be split into two non-contiguous regions, like Alaska and the mainland United States? Can an island be a country? Several islands? Can a country be entirely inside another country, like San Marino in Italy?

We need formal definitions, and in this case they get rather messy. Let's assume that each "country" is one contiguous region. Pick a "capital city" in each country. For each pair of neighboring countries, build a "road" to connect their capital cities in such a way that the road crosses a segment of the shared border and stays entirely within the area of the two countries. Do not allow two roads to cross each other. What you get is a *planar graph* — a graph drawn on a plane (Figure 17-11-a). The edges do not have to be straight line segments, but they cannot intersect.

The problem of coloring a map becomes the problem of coloring the vertices of the corresponding planar graph.

> **A graph is called *properly colored* if any two adjacent vertices are colored in different colors.**

It is not practical, of course, to "color" points; we simply assign them colors or numbers or symbols. Figure 17-11-b shows a graph properly colored in five colors, represented by the numbers 1 through 5.

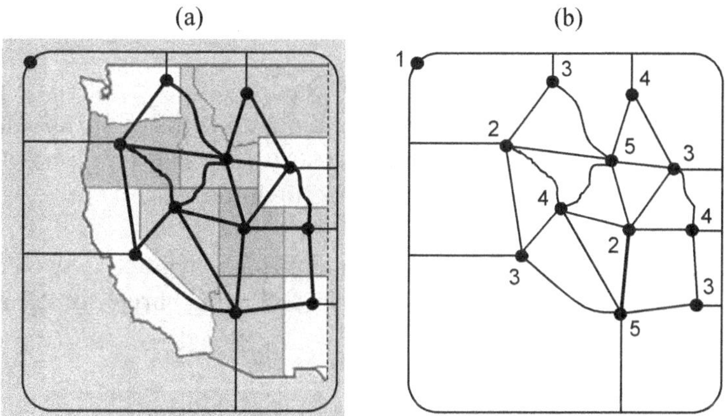

**Figure 17-11   (a) A planar graph corresponds to a geographical map
(b) The graph properly colored in five colors**

In coloring problems we consider only connected graphs, because if a graph is not connected, we can color each connected component separately.

Section 17.7 ~ Exercises

1.   Properly color the graph below in two colors. ✓

2.   Remove one vertex of degree 2 from the periphery of the graph in Question 1, merging the two edges that come out of it into one. Can the resulting graph be properly colored in two colors?

3.   Suppose a graph is properly colored in two colors. What can you tell about the colors of the two vertices that are the endpoints of a path of a certain length? When are they the same? When are they different? ✓

4.   Suppose a graph is properly colored in two colors, and A and B are two vertices connected by an edge. Consider any two paths from A and B, respectively, to a third vertex C. What can you tell about the parities of the lengths of these paths? Are they even or odd?

5.▪   Formulate the necessary and sufficient condition for a graph to be properly colorable in two colors. State your condition in terms of the absence of certain type of subgraphs in a graph. Justify your answer by showing that this is indeed a necessary and sufficient condition. ✓

6.◆   Write and test a Python function that colors a given graph in two colors or establishes that it can't be done. Use the following "brute-force" algorithm:

   1.   Color any one vertex.

   2.   For each vertex that is already colored, find its neighbors that are not yet colored. Assign each neighbor the appropriate color.

   3.   Repeat Step 2 until all the vertices are colored.

   4.   Check whether the graph is properly colored; return `None` if it isn't.

Assume that the input graph is represented by an n-by-n adjacency matrix, and return the result as a list of 1s and 2s that holds the colors of the vertices corresponding to the rows of the matrix.

7.   Properly color the graph in Figure 17-11-b in four colors.

8.   Consider the following graph:

Convince yourself that in any proper coloring of this graph in three colors, the colors of the vertices A and B are the same. Now, for any $N \geq 2$, give an example of a "longer" graph with similar properties:

a)   it is colorable in three colors;

b)   it has two vertices A and B such that the distance between them (the length of the shortest path) is greater than or equal to N;

c)   in any proper coloring of the graph in three colors, the colors of A and B are the same.

9.■   None of the following graphs is colorable in three colors:

Clearly, if you take any odd cycle and connect all its vertices to one "central" vertex, the resulting graph cannot be colored in three colors. Any graph that contains one of these "odd cartwheels" as a subgraph cannot be colored in three colors either. Come up with an example of a graph that does not contain an "odd cartwheel" and still is not colorable in three colors. ⸮ Hint: see Question 8. ⸗ ✓

10.✦   Give an example of a graph in three-dimensional space that cannot be colored in three colors and contains no "triangles" (that is cycles of length 3). ⸮ Hint: you will need at least eleven vertices. ⸗ ✓

17.8 The Four Color Theorem

The Four Color Theorem states that any planar map or graph can be properly colored using four colors. It first came up in the mid-19th century, but the proof evaded mathematicians for a long time after that. Many amateurs and professionals tried to show the theorem to be wrong by coming up with a counterexample: a map or a graph that cannot be colored in only four colors. They failed. Finally, the Four

Color Theorem was proved in the late 1970s by Kenneth Appel and Wolfgang Haken. Their proof, published in 1977, was unorthodox: they had to analyze many graph configurations, for which they used a computer program. It required 1200 hours of computer time to complete the proof. (These days it would take less time, of course.) Many mathematicians remained skeptical, though, because they could not verify the proof independently. In 1996 Neil Robertson, Daniel P. Sanders, Paul Seymour and Robin Thomas published a shorter and a more manageable proof, still based on Appel's and Haken's ideas.[*]

In this section we will make a naïve attempt to prove the theorem. We will largely follow the ideas of the British mathematician Alfred Kempe, who proposed his proof at the end of the 1870s. Kempe's proof stood unchallenged for about 10 years, until a major flaw was found in it. We'll go as far as we can with our proof and see what we can learn from it. At the end you yourself will discover and explain the flaw (see Question 10 in the exercises).

As often happens in proofs of theorems about graphs, we try to use mathematical induction. The idea is to somehow reduce a graph with n vertices to a smaller graph by eliminating one or several vertices. The smaller graph has the desired property by the induction hypothesis. We then try to restore the eliminated vertices in such a way that the property still applies. One has to be very careful doing all this (see Question 5 in the exercises).

One way to reduce the number of vertices in a graph is to "glue together" two vertices. If A and B are two vertices, we can replace them with one vertex O. We connect O to a vertex X with an edge if A <u>or</u> B (or both) is connected to X. (Question 2 in the exercises offers an example of "gluing together" two vertices.) We can glue together three or more vertices in the same way.

Another idea is to split a graph into smaller graphs, then, knowing that each of them has the desired property, combine them back into the original graph, while maintaining the property (see Question 3 in the exercises).

The edges of a planar graph divide the plane into regions, with one infinite outer region. If a graph is properly colored, and you remove one or several edges, the resulting graph will be properly colored, too. When proving theorems about coloring planar graphs, we can consider only the worst-case scenario, in which no edge can be added to the graph. This happens when all the regions are "triangles" (that is, are bounded by three edges). Such planar graphs are called *fully triangulated*. We can convert any planar graph into a fully triangulated graph by adding a few "diagonals"

---

[*] http://people.math.gatech.edu/~thomas/FC/fourcolor.html

to every region (see an example in Question 4 in the exercises). If we can color this fully triangulated graph in p colors, then we can color the original graph in p colors, too.

We are now done with the preliminaries and can proceed with our "proof." We will use mathematical induction by the number of vertices in the graph. Clearly any graph with four or fewer vertices can be colored in four colors (the base case). Let us take a graph with n vertices, $n > 4$. We assume (induction hypothesis) that any planar graph with fewer than n vertices can be colored in four colors and try to prove that our graph with n vertices can be colored in four colors, too.

Without loss of generality, we can assume that our graph is fully triangulated. Moreover, if any of the triangles has vertices both inside and outside, the problem of reducing the graph into smaller subgraphs is solved (see Question 3 in the exercises). So the neighborhood of any vertex looks like a simple "cartwheel" with the center at the vertex and at least three "spokes":

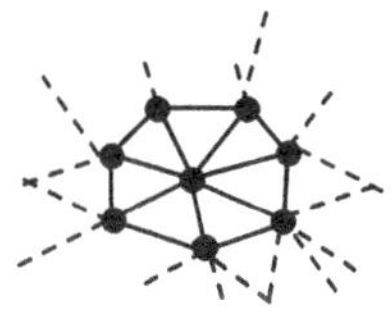

Let's take the vertex in the graph that has the smallest degree, vertex O. If the degree of O is 3, the problem is solved. Indeed, O is inside a triangle, and there can be no vertices outside (see Question 3 in the exercises). This means our graph is simply K_4:

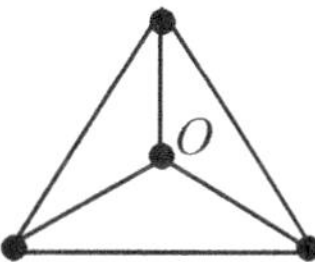

If the degree of O is 4, we need a little more work, but not much.

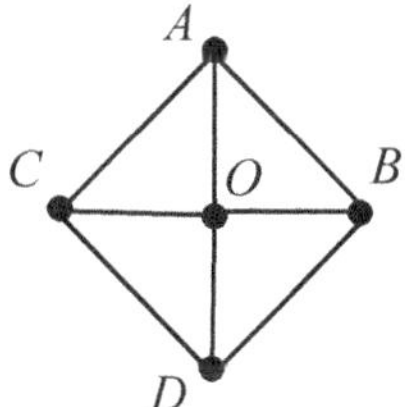

Let us "glue" the vertices B, O, and C together and properly color the resulting graph — we can do that by the induction hypothesis. When we unglue B, O, and C, all three will be colored in the same color, say color 1. The rest of the graph will be properly colored. Vertices A and D use at most two colors, say 2 and 3. Color 4 remains free, and we can recolor O in that color and get a proper coloring of our graph:

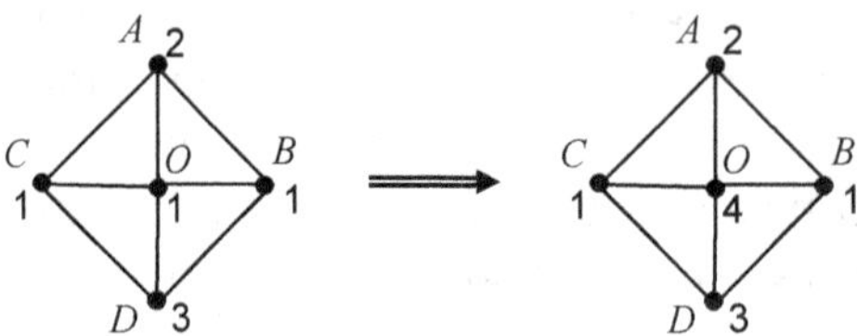

It turns out that any planar graph has a vertex of degree 5 or less (see Question 6 in the exercises). We have already considered the cases when the degree of O is 3 or 4. The only remaining case is when the degree of O is 5. This is the hardest case. If we remove O and the edges that connect it to its neighbors and add two "diagonals" to restore full triangulation, we get a smaller graph:

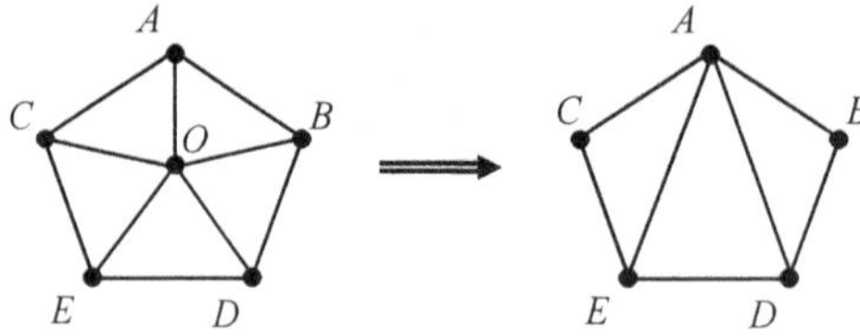

By the induction hypothesis, we can color this smaller graph in four colors. Unfortunately, A, B, C, D, and E, can use all four colors among them:

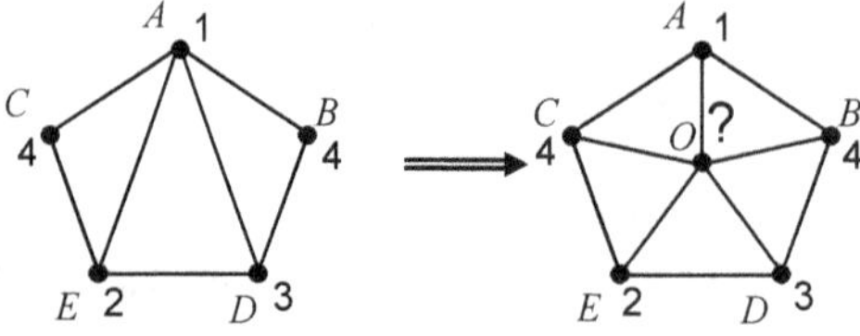

There is no easy way to simply recolor O and get proper coloring. Perhaps we can somehow recolor one of the vertices A, D, or E, or recolor both B and C to free one of the colors and use it for O. But how? That's where Kempe's idea, known as *Kempe's chains*, comes in.

A Kempe's chain is a path in a properly colored graph whose vertices are colored in two colors. Consider a subgraph in a properly-colored graph that consists of all the vertices colored in a pair of colors, say 1 and 2, and all the edges that connect them. This subgraph is not necessarily connected: not all of its vertices are necessarily connected to each other by a 1-2-colored "chain" (path). If this 1-2-colored subgraph is not connected, it splits into several connectivity components. We can take one of these components and flip the colors in it: 1 into 2 and 2 into 1. If our original graph was properly colored, the new coloring will be also proper. This method allows us to recolor some of the vertices in a properly colored graph without disturbing the other vertices.

Let us see how this applies to our proof. If A and E in the figure above are not connected by a 1-2 chain, we can take the largest connected 1-2 subgraph around A and flip the colors in it without disturbing the colors of B, C, D, and E. A will be recolored from 1 to 2, and 1 will be freed to color O. Similarly, if there is no 1-3 chain from A to D, we can recolor A in color 3 and free 1 for O.

So far so good. The only situation that remains to consider is when there are both a 1-2 chain from A to E and a 1-3 chain from A to D. For example:

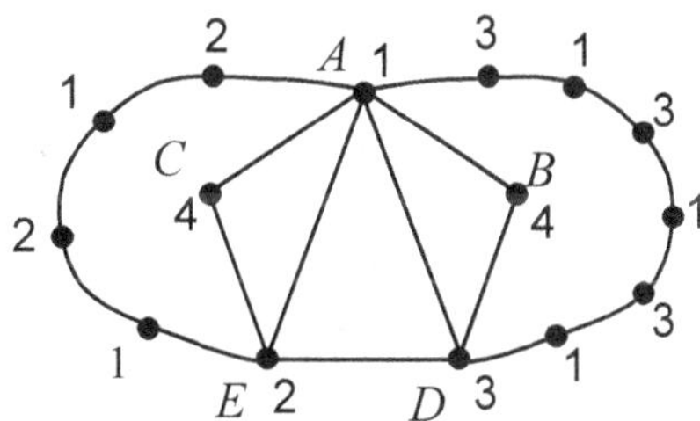

These chains serve as barriers on the plane between other pairs of vertices. The 1-2 chain from A to E ensures that there is no 4-3 chain from C to D, because 1-2 and 4-3 chains can't intersect. (If they did intersect in a vertex, what color would that vertex be?) So we can recolor C from 4 to 3. Likewise, we can recolor B from 4 to 2. 4 will be freed to color O.

This would complete our proof, if there weren't a big flaw, or, as mathematicians say, a "hole" in it. There is literally a hole in Kempe's chains: he didn't consider what "leaky" barriers they make. The 1-2 and 1-3 chains from A can take a roundabout path to their destinations instead of the direct path shown above. They can intersect. The 4-2 and 4-3 chains can "leak" out of their barriers and get "close" to each other, causing trouble: two vertices on these chains can be connected by an edge. Question 10 in the exercises asks you to provide a counterexample to Kempe's proof.

Section 17.8 ~ Exercises

1.   The map below has six regions (including the outer region). Draw a corresponding planar graph and properly color in four colors both the map and the graph.

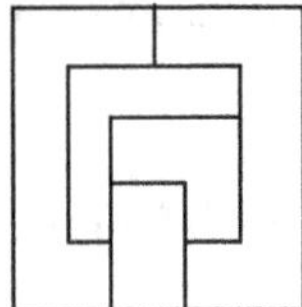

2.   Consider this graph:

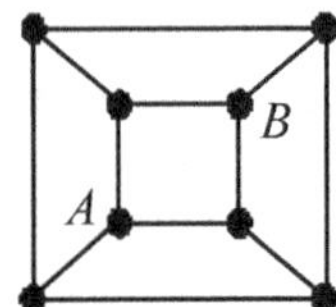

Draw the graph obtained from it by gluing A and B together. ✓

3.▪   Suppose we have a planar graph with n vertices. Suppose we know somehow (for example, from an induction hypothesis) that any planar graph with less than n vertices can be colored in p colors. Suppose also that our graph contains a "triangle" (a region bounded by three edges) with some vertices inside and some vertices outside. Prove that our graph can be colored in p colors, too. ✓

4.   Convert the following graph into a fully triangulated graph by adding edges (but not vertices):

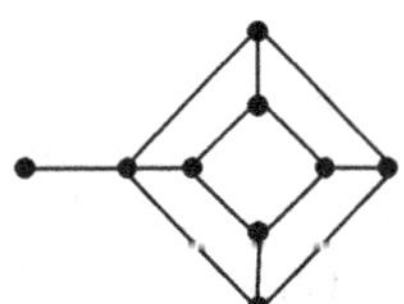

≷ Hint: don't forget the infinite outer region: it should be a "triangle," too. ≷ ✓

5. One has to be very careful with math induction proofs. Consider the following "proof" of an obviously incorrect statement: Any graph can be properly colored in three colors.

1. <u>Base case</u>. The statement is obviously true for a graph with 3 or fewer vertices.

2. <u>Inductive case</u>. Suppose the statement is true for any graph with less than n vertices (inductive hypothesis); let us prove that then the statement is true for any graph with n vertices. Let's take a graph with n vertices. Let's take any two vertices that are not connected by an edge and glue them together. By the induction hypothesis, we can color the resulting graph in three colors. Now let's unglue back the vertices , preserving the coloring. The vertices that were glued have the same color, but that's OK, since they are not connected by an edge. The original graph is now properly colored, too.

Find a flaw in this "proof." ✓

6. Prove that any planar graph has at least one vertex of degree 5 or lower. ⸰ Hints: it is sufficient to prove this for fully triangulated graphs; use Euler's formula that relates the number of vertices, edges, and regions in a planar graph: $V - E + R = 2$; estimate the number of edges in two ways: from the triangular regions and from the degrees of vertices. ⸰ ✓

7. Give an example of a fully triangulated planar graph such that all its vertices have a degree of 5 or higher. What is the smallest number of vertices of degree 5 in such a graph? ✓

8. In the graph in Figure 17-11, take the vertex in "Oregon," colored in color 2, find its 2-4 component, and flip the colors in it.

9. Suppose O is a vertex in a fully triangulated graph and its degree is greater than or equal to four. Show that there are at least two neighbors of O that are not connected by an edge.

10.◆   Complete the coloring of the graph below to make a counterexample to
Kempe's "proof."

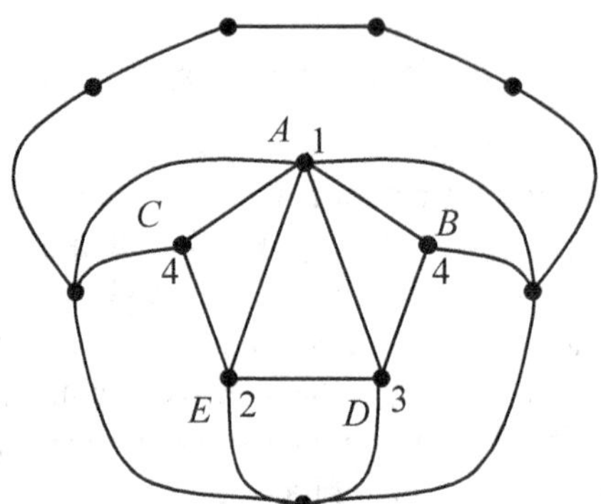

11.◆   Prove the Five Color Theorem: any planar graph can be properly colored in
five colors. ⟨ Hint: use the ideas from the "proof" of the Four Color
theorem. ⟩

17.9 Review

Terms and notation introduced in this chapter:

Graph	*Degree of a vertex*	$G = (V, E)$
Vertex or *node*	*Euler path*	K_n
Edge or *arc*	*Euler circuit*	
Multigraph	*Hamilton circuit*	C_n
Loop	*Directed graph*	
Simple graph	*Weighted graph*	
Complete graph	*Optimal path*	
Cycle	*Adjacency matrix*	
Isomorphism,	*Planar graph*	
isomorphic graphs	*Proper coloring*	
Equivalence relation	*Fully triangulated*	
Equivalence classes	*graph*	
Connected graph		

CHAPTER 18

Number Theory and Cryptology

18.1 Prologue

Number theory is a broadly defined branch of mathematics that deals with properties of numbers, especially integers. Many number theoretical concepts are easy to grasp: prime numbers, greatest common divisor, remainder, and so on. But some theorems are very difficult. Perhaps you have heard of Fermat's Last Theorem, which states that the equation $a^n + b^n = c^n$ has no positive integer solutions for $n > 2$. This theorem was proposed by Pierre de Fermat in the 1630s. Fermat left a note in the margin of his copy of Diophantus's *Arithmetica* stating that he had a marvelous proof of this theorem, but the margin was too narrow to write it down. Three and a half centuries of futile attempts to prove the theorem led to many advances in number theory. Finally, Andrew Wiles of Princeton University presented a proof in 1993. Wiles' proof relied on earlier results that linked Fermat's Last Theorem to the properties of a certain class of elliptic curves. The proof took over seven years to complete and three lectures to present.[*]

In this chapter we will only scratch the surface. We will start with Euclid's algorithm for finding the greatest common divisor, which will lead us to the fundamental theorem of arithmetic. We will take that opportunity to present a mini-theory, with definitions, theorems, and proofs, to give you a taste of such things. We will then consider arithmetic operations on remainders and applications of number theory to cryptology, the science of ciphers.

18.2 Euclid's Algorithm

Given two integers, a and $d \neq 0$, we say that a is evenly divisible by d (or, simply, that a is divisible by d) if $a = qd$ for some integer q. We can also say that d divides a or that d is a *divisor* of a.

> **We will consider only positive divisors: $d > 0$.**

> **The notation $d \mid a$ means d divides a.**

---

[*] It took Wiles another year and some help from his former student Richard Taylor to fill a gap discovered in his proof.

If $d \mid a$ and $d \mid b$, then d is a *common divisor* of a and b. GCD(a, b) stands for the greatest common divisor of a and b. (Sometimes, the greatest common divisor is called the *greatest common factor*, GCF.) Finding the GCD of two numbers is a common mathematical operation; it is used, for example, for reducing fractions.

You can find the GCD(a, b) for positive a and b by simply trying every number d, starting from 2 and up to a or b, whichever is smaller, and testing whether d is a common divisor. Such a "brute-force" approach, however, gets pretty tedious for big numbers.

Over 2300 years ago, in Book VII of his *Elements*, Euclid described an efficient algorithm for finding the GCD. Euclid's algorithm is based on the following key observation: if $b > a > 0$, then GCD(a, b) = GCD(a, $b - a$). (We leave the proof of this fact to you — see Question 1.) This allows us to go from a and b to smaller numbers, a and $b - a$. We repeat this operation several times, each time choosing the larger of a and b to reduce. Sooner or later we come to the situation where $a = b$. Then, of course, GCD(a, b) = a = b.

Example 1

Using Euclid's algorithm, find GCD(18, 30)

Solution

GCD(18, 30) = GCD(18, 12) = GCD(6, 12) = GCD(6, 6) = 6

Example 2

Write a Python function that uses Euclid's algorithm to find and return the greatest common divisor of two positive integers.

Solution

```python
def gcd(a, b):
    while a != b:
        if a > b:
            a -= b
        else:
            b -= a
    return a               # or return b
```

There is a much more efficient version of the `gcd` function that uses `a%b` instead of
`a - b` — see Question 4.

> **Two integers a and b are called *relatively prime* if they have no common
> divisors except 1, that is GCD(a, b) = 1.**

The concept of relatively prime numbers and the idea of Euclid's algorithm help us
analyze and solve equations of the type $ax + by = c$. Here a, b, and c are given
integers, and a solution is a pair of integers x and y. This type of equation is called a
linear Diophantine equation in two variables. In general, a polynomial equation in
one or several variables is called a *Diophantine equation* if we are looking only for
its integer solutions. Such equations are named after Diophantus, a Greek
mathematician who lived in Alexandria in the 3rd century and studied equations with
integer solutions. The first descriptions of linear Diophantine equations are found
much earlier, in Indian texts that are 2800 years old.

Example 3

A classroom has several desks. When we arrange the desks in rows of 5, all the rows
are complete, but when we arrange the desks in rows of 3, one desk remains:

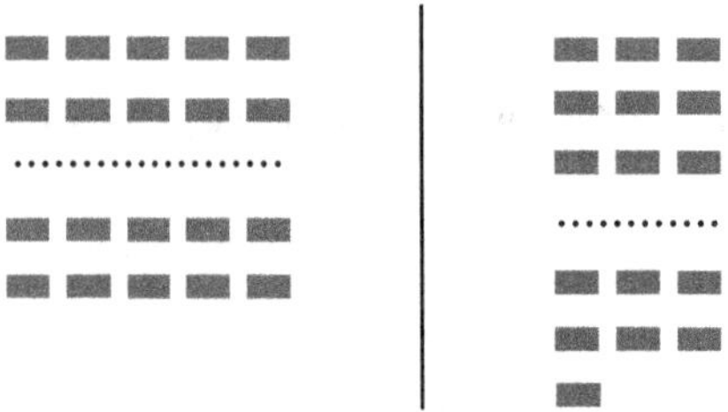

How many desks are there in the classroom, if their number is between 15 and 30?

Solution

If D is the number of desks, then $D = 5x$ and $D = 3y + 1$, where x and y are some
positive integers. Then $5x = 3y + 1$, which leads to the Diophantine equation
$5x - 3y = 1$.

This equation is easy to solve. We know that D is divisible by 5. We also know that
$D \geq 15$. We start counting by 5 from 15: 15, 20, ... — and soon get the answer:

$D = 25$. Indeed, when 25 is divided by 3, the remainder is 1. Here $x = 5$ and $y = 8$. If we keep going — 30, 35, ... — we soon get another number with the same properties, $D = 40$, but it is out of the given range.

Another way to solve this puzzle is to start from 1 and count by 3 until we get a number evenly divisible by 5 and within the 15-30 range: 1, 4, 7, 10, 13, 16, 19, 22, 25.

If the equation $ax + by = c$ has a solution, then it has infinitely many solutions, because if an x, y pair is a solution, then the $x + b$, $y - a$ pair is also a solution, the $x + 2b$, $y - 2a$ pair is a solution, and so on.

Does an equation $ax + by = 1$, where a and b are integers, always have a solution? Does $6x - 4y = 1$, for example, have a solution? Of course not! The left side must be an even number, so it can't be 1. What about $14x + 21y = 1$? Here $a = 14$, $b = 21$, and they have a common divisor 7. Therefore, the left side is always divisible by 7, so it can't be equal to 1. In general, if a and b have a common divisor greater than 1, then the equation $ax + by = 1$ has no solutions. What if a and b have no common divisors other than 1?

Linear Diophantine Equation Theorem

An equation $ax + by = 1$, where a and b are integers, has an integer solution if and only if a and b are relatively prime.

Proof:

We have already shown that if a and b have a common divisor greater than 1, then the equation has no solutions. Now let's assume that a and b <u>are</u> relatively prime and show that a solution exists.

Note that 0 is not relatively prime with any number, so we must have $a \neq 0$, $b \neq 0$. Without loss of generality we can assume that a and b are positive: if they are not, we can adjust the sign of x and/or y accordingly. For example, $ax + by = 1$ can be rewritten as $ax + (-b)(-y) = 1$. Also, a and b can't be equal, unless $a = b = 1$.

The main idea of the proof is the same as in Euclid's algorithm. Suppose $a > b$. Then $ax + by = 1$ can be rewritten as $(a - b)x + b(x + y) = 1$ or $(a - b)x_2 + by_2 = 1$. The original equation has a solution if and only if the new equation, with the coefficients $(a-b)$ and b, has a solution. $(a-b)$ and b are still positive and relatively prime. If we repeat this procedure several times, always subtracting the smaller coefficient from the larger one, we get smaller and smaller coefficients, until we come to $a=1$, $b=1$. Obviously the equation $x+y=1$ has solutions (for example, $x=1$, $y=0$ or $x=2$, $y=-1$). So the original equation must have a solution, too. Q.E.D.

Figure 18-1 illustrates the above proof for the equation $5x-3y=1$.

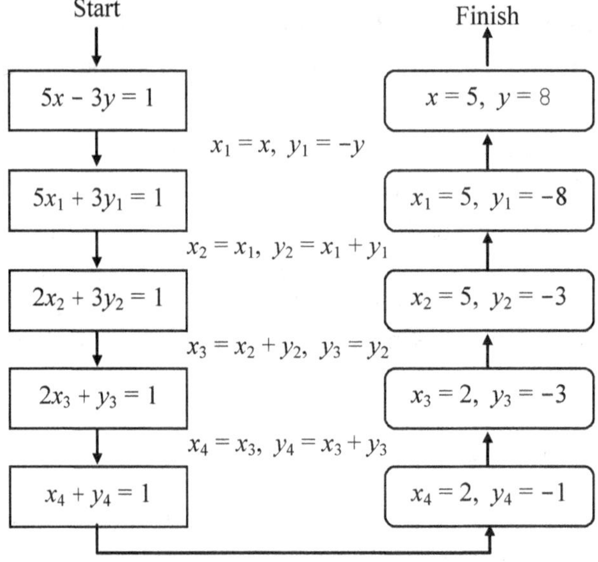

Figure 18-1. Finding a solution of a linear Diophantine equation by repeatedly reducing its coefficients

Section 18.2 ~ Exercises

1. Show that if $b > a > 0$, then GCD(a, b) = GCD($b - a$, a). ✓

2. Write a Python function `gcd(a, b)` that calculates the greatest common divisor of two positive integers a and b by using a "brute-force" approach: just check whether both a and b are divisible by d for all d from 2 to `min(a, b)`. ⸻ Hint: it is more efficient to go from `min(a, b)` down to 1. ⸻

3.   Rewrite the `gcd` function from Example 2 recursively, without loops. ✓

4.■   The `gcd` function in Example 2 can be made much more efficient. If $b > a$, instead of subtracting a from b multiple times, we can replace b with the remainder of the division of b by a (for example, using Python's `divmod` function). The same for $a > b$. Write and test this more efficient version of the `gcd` function. ⸮ Hint: make sure your code acts properly when a becomes divisible by b or b becomes divisible by a. ⸮ Compare the running times for the original and the modified code for $a = 18289894500228625200$, $b = 14814814692$. ✓

5.   Show that if a and b are relatively prime, then the equation $ax + by = c$ has an integer solution for any c. ✓

6.■   Show that $ax + by = c$ has an integer solution if and only if c is divisible by $GCD(a, b)$.

7.■   Write and test a function that finds a solution of the equation $ax + by = 1$ for given a and b and returns it as a tuple. Use the approach of Example 3.

8.■   Modify the function from Question 7 to find the solution of $ax + by = 1$ with the smallest possible positive integer x. Write another version of this function that returns the solution with x in a given range.

9.■   Write a function to solve $ax + by = c$ recursively. Use the approach outlined in the proof of the linear Diophantine equation theorem.

10.■   Show that if p and q are two different primes, any arithmetic sequence with the common difference p has a term that is divisible by q. ✓

11.■   Show that for any $k \geq 1$ there is no prime that divides all Fibonacci numbers starting from F_k.

12.   Is it true that if $GCD(a, b) = 1$ or $GCD(b, c) = 1$ or $GCD(c, a) = 1$, then the greatest common factor of a, b and c is 1? If so, prove it; if not, give a counterexample. Is the converse true? If so, prove it; if not, give a counterexample.

13.◆   Formulate and prove the necessary and sufficient condition for the equation $ax + by + cz = 1$ to have an integer solution. ✓

18.3 The Fundamental Theorem of Arithmetic

In this section we present a mini-theory that leads to the fundamental theorem of arithmetic. We present it in the typical mathematical style with definitions and theorems. To make our theory complete, we restate here the definitions from Section 18.2.

Definition 1:

Let n be an integer. A positive integer d is called its *divisor* if there exists an integer q such that $n = qd$.

Definition 2:

An integer p is called a *prime* if $p > 1$ and p has no divisors other than 1 and p.

Definition 3:

Two integers m and n are called *relatively prime* if they have no common divisors greater than 1.

Theorem 1:

Any integer greater than 1 is either a prime or can be represented as a product of primes.

Proof:

This is a proof by mathematical induction.

Base case: The statement is true for $n = 2$, because 2 is a prime.

Induction hypothesis: Suppose that the statement is true for any integer greater than 1 and less than n. Now let's show that the statement is then true for n. Indeed, if n is a prime, the statement is true for n. If n is <u>not</u> a prime, then $n = qd$, where $1 < d < n$ and $1 < q < n$. By the induction hypothesis, d is a prime or a product of primes. The same for q. Combining the products for q and d into one product we conclude that n is a product of primes, too. By mathematical induction, the statement is true for any $n \geq 2$, Q.E.D.

A slightly different take on this proof relies on the fact that any non-empty set of positive integers has a smallest number. (This fact is equivalent to math induction.) Consider the set of all numbers greater than 1 for which the statement is <u>not</u> true. If this set is empty, our proof is complete. If not, let n be its smallest element. n can't be a prime, so $n = qd$, where $1 < d < n$ and $1 < q < n$. Since n is the smallest number for which the statement is false, the statement must be true for q and d. But if we factor both d and q into primes, and combine them into one product, we get n. So the statement must also be true for n. This is a contradiction, so no such n can exist.

 **A *corollary* is a mathematical fact that easily follows from a theorem.**

<u>Corollary:</u>

There are infinitely many primes.

Proof:

Suppose the set of all primes were finite: $\{p_1, p_2, ..., p_N\}$. Let us consider $m = p_1 \cdot p_2 \cdot ... \cdot p_N + 1$. By Theorem 1, m must be a prime or have a prime divisor. That prime must be different from any of the primes $p_1, p_2, ..., p_N$ because none of them is a divisor of m. This contradicts the assumption that $\{p_1, p_2, ..., p_N\}$ is the set of all available primes.

<u>Theorem 2:</u>

If integers a and b are relatively prime, then there exist integers x and y such that $ax + by = 1$.

Proof:

This is a part of the linear Diophantine equation theorem from Section 18.2.

Theorem 3:

If p is a prime and $p \mid mn$, then $p \mid m$ or $p \mid n$.

This is Proposition 30 in Book VII of Euclid's *Elements*. It is also known as *Euclid's First Theorem*. The statement seems completely obvious, but its proof is not obvious at all. This happens often in mathematics.

Proof:

This is a proof by contradiction. Suppose p divides neither m nor n. Then p and n are relatively prime (because p is a prime). By Theorem 2, there exist integers x and y such that $px + ny = 1$. Multiplying both sides by m we get $mpx + mny = m$. The left side is divisible by p, because $p \mid mn$. So m must be divisible by p, too. Therefore, our assumption that p divides neither m nor n was false. Q.E.D.

Corollary:

If p is a prime and $p \mid n_1 \cdot ... \cdot n_k$, then $p \mid n_1$ or $p \mid n_2$ ... or $p \mid n_k$.

Proof:

By Theorem 3, $p \mid n_1$ or $p \mid n_2 \cdot ... \cdot n_k$. And so on.

❖   ❖   ❖

We are now ready to tackle the fundamental theorem of arithmetic.

Theorem 4 (the fundamental theorem of arithmetic):

Any integer greater than 1 can be represented as a product of primes, and such factorization is unique (if we disregard the order of factors).

Proof:

By Theorem 1, any integer greater than 1 is either a prime or a product of primes. Now we have to show that such factorization is unique. Suppose there exist numbers with two different factorizations. Let's take the smallest such number $n = p_1 \cdot ... \cdot p_i = q_1 \cdot ... \cdot q_j$. $q_1 \mid n$, so $q_1 \mid p_1 \cdot ... \cdot p_i$. By Theorem 3, q_1 must be one of the primes $p_1, ..., p_i$. Therefore, we can divide $p_1 \cdot ... \cdot p_i$ and $q_1 \cdot ... \cdot q_j$ by q_1 and get a smaller number n / q_1 with two different factorizations. This is a contradiction, so no such n can exist. Q.E.D.

Section 18.3 ~ Exercises

1. Prove that for any positive integer k there exist k consecutive positive integers such that none of them is a prime. ⸨ Hint: start at $(k+1)! + 2$. ⸩

2. 7, 13, 19 form an arithmetic sequence and are all primes. Show that an infinite arithmetic sequence cannot contain only primes. ⸨ Hint: see Question 1. ⸩

3.■ Write and test a program that prompts the user to enter an integer n greater than 1 and prints all its prime factors in ascending order. Each factor must be printed as many times as it appears in the factorization of n. The dialog with the program should look like this:

    ```
    Enter an integer greater than 1: 90
    90 = 2*3*3*5
    ```

    or

    ```
    Enter an integer greater than 1: 3
    3 = 3
    ```

4. If $p_1, p_2, ..., p_n$ are the first n primes, is it always true that $p_1 \cdot p_2 \cdot .... \cdot p_n + 1$ is a prime? If you believe it is true, prove it. If not, find a counterexample. ✓

5. Write a program to find the smallest positive n such that $n^2 - n + 41$ is <u>not</u> a prime. (There is no polynomial, even in several variables, that produces only prime values.)

6. Show that any prime greater than 2 can be uniquely represented as $a^2 - b^2$, where a and b are positive integers. ✓

7.■ Write a program to find the biggest prime among the first 100 Fibonacci numbers. (Mathematicians don't yet know whether there are infinitely many primes among Fibonacci numbers.) ✓

8. *Goldbach's conjecture* states that every even integer greater than 2 can be represented as the sum of two primes. For example, $12 = 5 + 7$. It is not known whether this conjecture is true or false. Write a program to verify Goldbach's conjecture for all even numbers from 4 to 100.

9. Show that the number of different divisors of n is even, unless n is a perfect square. The number of different divisors of a perfect square is odd.

10. Suppose $n = p_1^{j_1} \cdot ... \cdot p_k^{j_k}$, where $p_1, ..., p_k$ are different primes. Express the total number of divisors of n in terms of $j_1, ..., j_k$. ✓

11. Suppose $n = p^{j_1} \cdot q^{j_2}$, where p and q are two different primes. Show that the number of positive integers that are less than n and are relatively prime with n is $n\left(1 - \dfrac{1}{p}\right)\left(1 - \dfrac{1}{q}\right)$. Extend the result for three primes; for any number of primes. ✓

18.4 Arithmetic of Remainders

Let n and d be integers and $d > 0$. We can divide n by d with a remainder. This means that we can find integers q and r, such that $n = qd + r$, where $0 \le r < d$. Here q is the quotient and r is the remainder.

> **Two integers m and n are said to be *congruent modulo d*, if they give the same remainder when divided by d. This is written as $m \equiv n \pmod{d}$.**
>
> **n mod d denotes the remainder when n is divided by d.**

For example, $12 \equiv 27 \pmod 5$: both 12 and 27 give the same remainder when divided by 5: $12 \bmod 5 = 27 \bmod 5 = 2$.

If $m \equiv n \pmod d$, then $d \mid (m - n)$, that is, $m - n$ is divisible by d.

Congruence modulo d is an *equivalence relation* on integers (see Section 17.3). Indeed, all three required criteria for an equivalence relation are satisfied:

1. Reflexivity: $n \equiv n \pmod{d}$
2. Symmetry: If $m \equiv n \pmod{d}$, then $n \equiv m \pmod{d}$
3. Transitivity: If $k \equiv m \pmod{d}$ and $m \equiv n \pmod{d}$, then $k \equiv n \pmod{d}$

Therefore, for a given d, all integers fall into non-overlapping equivalence classes with respect to congruence modulo d. Two integers from the same class give the same remainder when divided by d, and two integers from different classes give different remainders. The number of congruence classes is d.

Example 1

List the congruence classes modulo 3.

Solution

There are three classes:

$$\{... -12, -9, -6, -3, 0, 3, 6, 9, 12, ...\} \equiv 0 \pmod{3}$$
$$\{... -11, -8, -5, -2, 1, 4, 7, 10, 13, ...\} \equiv 1 \pmod{3}$$
$$\{... -10, -7, -4, -1, 2, 5, 8, 11, 14, ...\} \equiv 2 \pmod{3}$$

❖ ❖ ❖

If we take any integer k not divisible by 3 and add it to all numbers in a congruence class modulo d, all of them will shift into another congruence class. For example, –4, 5, and 14 all belong to the "2" congruence class modulo 3. If we add 7 to each of them, we get 3, 12, and 21, all of which belong to the "0" congruence class. This is so, because $(n+k) \bmod d$ and $(n \bmod d)+(k \bmod d)$ are congruent modulo d:

$$(n+k) \bmod d \equiv \left((n \bmod d)+(k \bmod d)\right) \pmod{d}$$

The same is true for multiplication:

$$(n \cdot k) \bmod d \equiv \left((n \bmod d)\cdot(k \bmod d)\right) \pmod{d}$$

All of this means that we can add and multiply two congruence classes, or more precisely, any two representatives of these congruence classes. The result falls into the same class, regardless of which representatives we chose.

From now on, we will represent each congruence class modulo d simply by all possible remainders when n is divided by d: $0, 1, 2, ..., d-1$. Let's use the symbol $\oplus_d$ to denote addition modulo d. The sum of two numbers modulo d simply "wraps around" at d. It is as if the number line was wrapped around a circle with the numbers $0, 1, 2, ..., d-1$ evenly spaced on it (Figure 18-2). We are all familiar with such an arrangement on a clock, which adds hours modulo 12 (or modulo 24 on a European digital clock) and adds minutes modulo 60. The only difference is that in math we usually go counterclockwise.

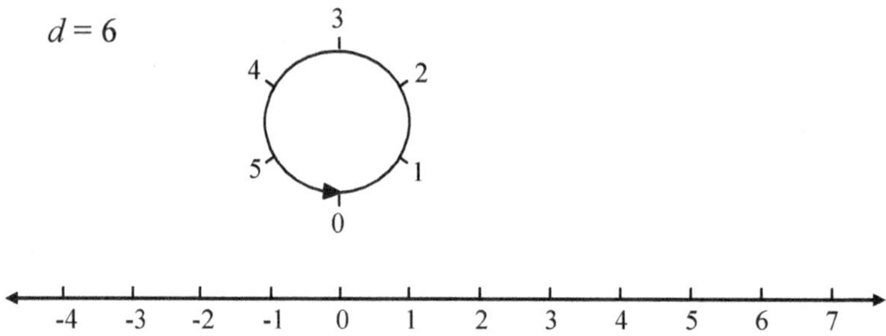

Figure 18-2.   Addition and multiplication modulo d can be viewed as performed on a "number circle"

Example 2

$2 \oplus_3 2 = 4 \mod 3 = 1$, is in the same congruence class as $5 \oplus_3 11 = 16 \mod 3 = 1$.

Example 3

$7 \oplus_{12} 10 = 17 \mod 12 = 5$

The same principle is used for multiplication. We will use the $\otimes_d$ symbol to denote multiplication modulo d.

Example 4

$2 \otimes_3 2 = 4 \ \mathrm{mod}\ 3 = 1$

$3 \otimes_5 4 = 12 \ \mathrm{mod}\ 5 = 2$

$7 \otimes_{12} 10 = 70 \ \mathrm{mod}\ 12 = 10$

Figure 18-3 shows modulo 5 addition and multiplication tables.

$\oplus_5$	0	1	2	3	4		$\otimes_5$	0	1	2	3	4
0	0	1	2	3	4		0	0	0	0	0	0
1	1	2	3	4	0		1	0	1	2	3	4
2	2	3	4	0	1		2	0	2	4	1	3
3	3	4	0	1	2		3	0	3	1	4	2
4	4	0	1	2	3		4	0	4	3	2	1

Figure 18-3. Modulo 5 addition and multiplication tables

Modulo d arithmetic is similar in many ways to arithmetic on integers:

- It obeys the associative and commutative laws for addition and subtraction;
- It obeys the distributive law;
- There is a 0;
- There is a 1;
- Each element x has an additive inverse element y, such that $x \oplus_d y = 0$ (namely, $y = d - x$).

But there is a peculiarity, too: if d is <u>not</u> a prime, then you can find x and y such that $x \neq 0$, $y \neq 0$, but $x \otimes_d y = 0$. For example, $2 \otimes_6 3 = 0$. Such x and y are called *divisors of zero*.

If p is a prime, then multiplication modulo p has no divisors of zero. Moreover, for each $x \neq 0$ there is a y such that $x \otimes_p y = 1$. For example, take a look at the multiplication table in Figure 18-3: $2 \otimes_5 3 = 1$. Every row of the table has a 1. We can say that in modulo p arithmetic, $y = \dfrac{1}{x}$. We can do division! So the arithmetic of remainders modulo p, for a prime p, is similar to the arithmetic of rational numbers.

Let's prove this fact: that in modulo p arithmetic any $x \neq 0$ has a reciprocal $y = \dfrac{1}{x}$.

But first we need to review a couple of properties of exponents:

$$x^0 = 1$$
$$x^m \cdot x^n = x^{m+n}$$

Theorem 1:

Let p be a prime. For any $0 < x < p$, there exist a $0 < y < p$, such that $xy \equiv 1 \ (\text{mod } p)$.

Proof:

Consider the geometric sequence $1, x, x^2, x^3, \dots$. Since there are only p different remainders modulo p, sooner or later two terms in this sequence will have the same remainder. Suppose $x^n \equiv x^m \ (\text{mod } p)$ where $n > m$. Then $x^n - x^m \equiv 0 \ (\text{mod } p)$. $x^n - x^m = x^m(x^{n-m} - 1)$ and p does not divide x^m. Then, by Theorem 3 in Section 18.3, p divides $x^{n-m} - 1$. So $x^{n-m} \equiv 1 \ (\text{mod } p)$. We can take $y = x^{n-m-1}$. Q.E.D.

❖ ❖ ❖

If p is a prime and we take any $0 < a < p$ and look at the remainders of $1, a, a^2, a^3, \dots$ modulo p, we can see that they form a periodic sequence:

mod 7	1	a	a^2	a^3	a^4	a^5	a^6	a^7
$a = 2$	1	2	4	1	2	4	1	2
$a = 3$	1	3	2	6	4	5	1	3
$a = 4$	1	4	2	1	4	2	1	4

The period of the sequence always divides $p - 1$, and $a^{p-1} \equiv 1 \ (\text{mod } p)$.

<u>Theorem 2 (Fermat's Little Theorem):</u>

For any positive integer a and a prime p, $a^p \equiv a \pmod{p}$.

Proof:

If $p \mid a$, then $a^p \equiv a \equiv 0 \pmod{p}$. If not, as we saw above, $a^{p-1} \equiv 1 \pmod{p}$ $\Rightarrow$ $a^p \equiv a \pmod{p}$.

Section 18.4 ~ Exercises

1.   Explore what Python's `n%d` operator returns when $n < 0$ and/or $d < 0$.

2.   Show that addition and multiplication modulo 2 corresponds to the XOR and AND operations, respectively, where FALSE is 0 and TRUE is 1.

3.   Write and test a function `elapsed_time` that returns the time difference in minutes from `(hour1, min1)` to `(hour2, min2)`. ✓

4.■   Suppose the days of the week are represented by numbers: Sunday — 0, Monday — 1, and so on. Write and test a function that takes the day of the week for January 1 and returns the date of Thanksgiving (fourth Thursday in November) for that year (assuming this is not a leap year). ✓

5.   Write by hand the addition and multiplication tables modulo 6. ✓

6.   Show that there are two numbers in any set of 101 integers (not necessarily consecutive) whose difference ends with two zeros.

7.   Show that if p is a prime and $p \mid a^2$ then $p^2 \mid a^2$.

8.■   Find the smallest positive n such that $n \equiv 2 \pmod{3}$, $n \equiv 3 \pmod{4}$, $n \equiv 4 \pmod{5}$, and so on up through $n \equiv 11 \pmod{12}$. ✓

9.■   Calculate $3^{22222} \bmod 23$ using pencil and paper only. ✓

10.■   Show that for any positive n, $6 \mid n(n+1)(2n+1)$.

11.■ Show that if p is a prime greater than or equal to 5, then $p^2 - 1$ is evenly divisible by 24. ✓

12.■ Write and test a function that takes a positive integer a and a prime p and returns the smallest $k > 1$ such that $a^k - a \equiv 0 \pmod{p}$.

13. Show that if p and q are two different primes, and $0 \le a < p$ and $0 \le b < q$, you can find x such that $x \equiv a \pmod{p}$ and $x \equiv b \pmod{q}$. (This is a simple special case of the *Chinese remainder theorem*.) ≋ Hint: See Question 10 in Section 18.2. ≋

14. Wilson's theorem states that if p is a prime, $(p-1)! + 1$ is evenly divisible by p.

   (a) Show that Wilson's theorem is true only for primes.

   (b)◆ Prove Wilson's theorem. ≋ Hint: recall that every x has a reciprocal $y = \dfrac{1}{x}$ modulo p, and that $x = y$ only for $x = 1$ and $x = p - 1$. ≋ ✓

   (c)■ Write and test a program that prints out the first 100 primes using Wilson's theorem.

15.◆ Fermat's Little Theorem can be used to test a number p for primeness: p is a prime if and only if $a^{p-1} \equiv 1 \pmod{p}$ for all positive a that are not divisible by p. We can choose several such values of a at random and test the condition; if it works for all of them, then the probability of p being a prime is very high. Write a function that takes a positive integer p and checks whether it is likely to be a prime by using up to 100 random values for a ($1 < a < p$). Using this function, find the first prime over 1,000,000.

   ≋ Hints:

   1. First write a function that quickly computes $a^n \bmod p$ with intermediate numbers not exceeding ap.

   2. Recall that the `randint(a, b)` function from the `random` module returns a random integer from `a` to `b` (inclusive).

   ≋

18.5 Ciphers

Cryptology is the science of making and analyzing ciphers. In a simple case, if Alice and Bob want to send each other encrypted messages, all they have to do is agree on which encryption method they are going to use. ("Alice" and "Bob" are often used in cryptology literature to explain secure communications between two people or organizations. There is a third character, the eavesdropper "Eve," who monitors all exchanges between Alice and Bob. Eve will try to guess which method Alice and Bob are using.)

Example 1

The simplest substitution cipher: each letter is replaced with the next letter in the alphabet; 'Z' is replaced by 'A'. "Got an A" is encrypted as "Hpu bo B".

This type of cipher is easy to break once many people start using the same method. In a more advanced cipher, the encryption method is widely known, but Alice and Bob share a secret *key*, which modifies the encryption scheme.

Example 2

In a simple substitution cipher, a secret key tells which letter should replace 'A', 'B', ..., 'Z'. For example:

```
Letter to encrypt:    ABCDEFGHIJKLMNOPQRSTUVWXYZ
Key:                  QEKUOYMBJXRCDZNVTGFASHWPLI
```

"On time" is encrypted as "Nz ajdo." This cipher is easy to break by comparing the frequencies of occurrence of different letters in the plain text and encoded text and by guessing some common short words (articles, prepositions, and so on).

In a cipher that uses a secret key, Alice and Bob must somehow share the key. They can meet in person or send a sealed envelope by snail mail. This might work if A and B are just two people. But what if A is amazon.com and B is any customer who wants to place an order in a secure manner? Imagine 100,000,000 Internet customers and 100,000 secure servers at e-commerce sites. You would need to distribute 10,000,000,000,000 different keys to allow each to communicate with each! To address this problem, companies use several very clever schemes that emerged in the 1970s. We will consider two of them: *Diffie-Hellman Key Exchange* and *RSA public/private key encryption*.

The Diffie-Hellman Scheme

The Diffie-Hellman Key Exchange algorithm was invented by Whitfield Diffie and Martin Hellman. The key idea (pun intended!) of the D-H scheme is to let each member of a community own "one half" of a key. All these half-keys are public, either published in a directory or available upon request from the owner. Any two halves can be put together to make a unique key. However, the person combining the two halves must know the secret code for at least one of the halves to be able to join them together. Alice has the secret code to her half and Bob has the secret code to his half, so either Alice or Bob can put together their halves to make a complete key; the resulting key will be the same. But Eve cannot put together Alice's and Bob's half-keys.

In the numerical model of this scheme, Alice's public half-key is $K_A = r^a \bmod p$, Bob's public half-key is $K_B = r^b \bmod p$, and the combined key is $K = r^{ab} \bmod p$, where p and r are in the public domain and are used by everyone. p is a large prime: it is chosen to have about 200 digits! $1 < r < p$ (Figure 18-4).

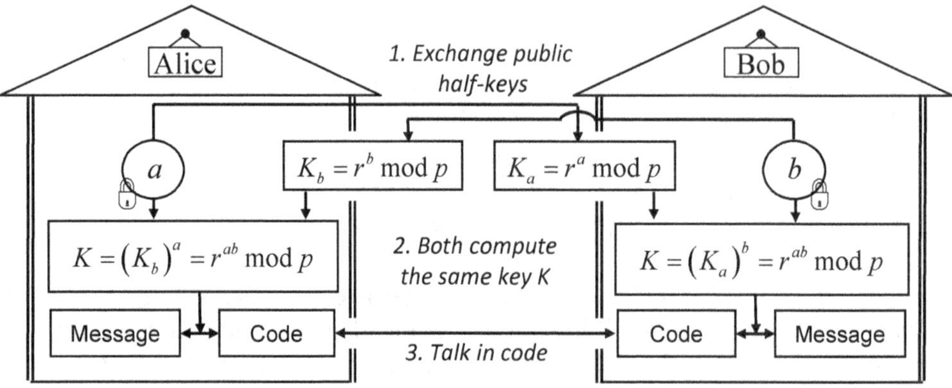

Figure 18-4. A metaphor for the D-H scheme: Alice's and Bob's half-keys are publicly available, but you need to know either Alice's or Bob's secret code to combine them into a complete key suitable for encoding and decoding messages.

In modulo p arithmetic, it is virtually impossible to calculate K from K_A and K_B for a very large p, unless you know a or b. It is also virtually impossible to find a from K_A or b from K_B. But if you know a or b, combining K_A and K_B into one key K is easy. Recall the property of exponents: $\left(r^a\right)^b = r^{ab}$ (see Question 3). Likewise, $\left(r^b\right)^a = r^{ab}$. So $K = \left(K_B\right)^a = \left(K_A\right)^b$.

Here are the steps Alice takes to send a coded message to Bob:

1. Alice knows the standard p and r from the public domain.

2. Alice requests K_B from Bob.

3. Alice makes the encryption key $K = \left(K_B\right)^a \bmod p = r^{ab} \bmod p$. using her secret number a.

4. Alice encodes the plain text message M into the encoded message C using K as the key for encryption: $C = \text{encode}(M, K)$. (It is assumed that everyone is using the same encryption method — only the keys are different.)

5. Alice sends C to Bob.

Bob takes similar steps to decode the message received from Alice:

1. Bob knows the standard p and r from the public domain.

2. Bob requests K_A from Alice.

3. Bob makes the encryption key $K = \left(K_A\right)^b \bmod p = r^{ab} \bmod p$ using his secret number b.

4. Bob decodes C using K: $M = \text{decode}(C, K)$.

In regular arithmetic, if you know r and r^a, it is very easy to calculate a. For example, if $r = 10$, and $r^a = 1000$, you know right away that $a = 3$.

If $r^a = x$, then a is called the *logarithm of x to the base r*. It is written as $a = \log_r x$.

In regular arithmetic, $\log_r x$ increases in a predictable way when x increases, which makes it easy to compute. Not so in modulo p arithmetic. When a changes from 1 to $p-1$, $r^a \bmod p$ jumps around in the range from 1 to $p-1$.

 $a = \log_r x \pmod p$ **is called the *discrete logarithm*.**

Example 3

Find $\log_6 8 \pmod{11}$, that is, find a such that $6^a \equiv 8 \pmod{11}$.

Solution

Make a table of powers of 6 modulo 11:

a	0	1	2	3	4	5	6	7	8	9	10
$6^a \bmod 11$	1	6	3	7	9	10	5	8	4	2	1

An entry in the bottom row is the previous entry times 6 (mod 11). For example $6 \cdot 6 = 36 \equiv 3 \pmod{11}$; $3 \cdot 6 = 18 \equiv 7 \pmod{11}$. Look in the bottom row for the power of 6 that is equal to 8 and mark the corresponding a — here it is $a = 7$. Therefore, $\log_6 8 \pmod{11} = 7$.

Can't we make and search a similar table for any r and p? Not if p has 200 digits! We can always find r such that all the powers of r modulo p are different. Then the table of powers of r will have about 10^{200} columns! This is more than the number of atoms in the universe... squared! If we combine all the computers in the world (including all smartphones[*]) and make them search for a discrete logarithm, it will take more time than the age of the universe... cubed!

---

[*] In 2007, The Guinness Book of Records recognized folding@home (FAH) as the world's most powerful distributed computing network. FAH had signed up more than 670,000 PS3 gaming consoles to analyze, during their idle hours, the shapes of proteins and their effects on various diseases. These days the network runs on smartphones and is estimated to perform at over 100 petaFLOPs (10^{17} floating-point operations per second).

> **The discrete logarithm is virtually impossible to compute for very large p.**

On the other hand, calculating r^a is relatively quick, even for a very large a. For example, you need to perform only 10 multiplications to calculate r^{1024} (see Question 9 in Section 16.3).

The RSA Algorithm

The name of the RSA algorithm is formed by the last name initials of its inventors, Ron Rivest, Adi Shamir, and Leonard Adleman at MIT. They invented the algorithm in 1977, independently of previous work.[*] The RSA algorithm is now widely used for secure communications and e-commerce on the Internet.

The main idea of the RSA scheme is simple. Suppose Alice wants to send a secret message to Bob. Alice asks Bob to send her an open padlock to which only Bob has the key. (Bob, like every other member of the community, has an unlimited supply of such locks and sends them out for free on request.) When Alice receives the lock, she puts her message in a box, puts the lock on it, clicks it locked, and sends the box to Bob.

In the numerical implementation, Bob sends to Alice, over an open channel, two numbers, E and N. These numbers together serve as the "lock" (but they are called Bob's public encryption key). Alice's message is stored in a computer as a sequence of 0s and 1s, and the RSA algorithm treats these bits as binary digits of a positive integer M. Alice "puts a lock" on M, that is, creates a coded message $C = M^E \bmod N$ and sends C back to Bob. Bob uses his secret private key D to decode C and get back M: $M = C^D \bmod N$.

For this to work, we must have $C^D \equiv \left(M^E\right)^D \equiv M^{ED} \equiv M \pmod{N}$, for any positive integer M. How is this possible?

Recall Fermat's Little Theorem (Section 18.4), which states that if p is a prime, then $x^p \equiv x \pmod{p}$, for any positive integer x. This is our starting point. If x is relatively prime with p, then $x^{p-1} \equiv 1 \pmod{p}$, which means $p \mid (x^{p-1} - 1)$.

---

[*] British mathematician Clifford Cocks, who worked for a British intelligence agency, had described a similar algorithm in 1973 in a top-secret internal paper. His discovery remained unknown until 1997.

The RSA scheme relies on a more general theorem. In RSA, N is not a prime but a product of two different primes: $N = p \cdot q$. p and q are chosen to be very large primes, at least 100 digits each, so N has at least 200 digits (over 640 binary digits). It is easy to calculate $N = p \cdot q$ but it is virtually impossible to factor N if p and q are kept secret — it would take forever.

The math needed to get $M^{ED} \equiv M \pmod{N}$ is a little long but elegant:

1.   We limit M to $M < p$ and $M < q$. Then M is relatively prime to p and to q. So is any power of M.

2.   Applying Fermat's Little Theorem to $x = M^{y(q-1)}$ and p, where y is any positive integer, we get $p \mid \left(\left(M^{y(q-1)} \right)^{p-1} - 1 \right) \Rightarrow$

$p \mid (M^{y(q-1)(p-1)} - 1)$. Similarly, $q \mid (M^{y(p-1)(q-1)} - 1)$.

3.   Since both p and q are divisors of $M^{y(p-1)(q-1)} - 1$, their product N is also a divisor: $N \mid (M^{y(p-1)(q-1)} - 1)$. Therefore

$M^{y(p-1)(q-1)} \equiv 1 \pmod{N}$. Multiplying both sides by M we get

$M^{y(p-1)(q-1)+1} \equiv M \pmod{N}$.

4.   All of the above are well-known results in number theory. Now all we need is to represent $y(p-1)(q-1)+1$ as a product ED. Bob chooses an E such that it is relatively prime with $p-1$ and $q-1$. Then E is relatively prime with $(p-1)(q-1)$. Solving the Diophantine equation $Ex - (p-1)(q-1)y = 1$ (see Section 18.2), we find x and y and set $D = x$. Then $y(p-1)(q-1)+1 = ED$.

5.   We have found D and E, such that $M^{ED} \equiv M \pmod{N}$ for any M relatively prime with N (in particular, for any M such that $M < p$ and $M < q$).

Eve knows only $N = pq$ and E, but not p and q. She can't factor N or compute $(p-1)(q-1)$, so she can't compute D.

The RSA scheme is a little slow, so, in practice, it is used only to send a secret key for a different cipher. Once both Alice and Bob know that secret key, they can communicate in that cipher.

Section 18.5 ~ Exercises

1. Write and test two functions, `encode(text, key)` and `decode(code, key)`, that implement a substitution cipher with a key. `text`, `code`, and `key` are strings, and each of the functions returns a string.

2. Look online for a description of the Vigenère cipher. Write and test functions `encode(text, key)` and `decode(code, key)` that use that cipher. ✓

3. Show that for positive integers a and b, $\left(r^a\right)^b = r^{ab}$.

4. $p = 170141183460469231731687303715884105727$ is a prime.
$a = 618970019642690137449562111$.
$r = 5$.
Calculate $r^a \bmod p$. ⟨ Hint: use the $r^a \bmod p$ function from Question 15 in Section 18.4. ⟩ ✓

5. Show that if $x \equiv a \pmod{p}$ and $x \equiv a \pmod{q}$, then $x \equiv a \pmod{pq}$.

6. In the RSA algorithm, given $p = 13$, $q = 17$, and $E = 5$, find D. ✓

7. In describing RSA, we used a metaphor of Bob sending an open lock to Alice and Alice sending Bob a secret message in a box locked by that lock. Suppose only locked boxes are allowed in the mail. Can Alice still send a secret message to Bob? ⟨ Hint: two locks can fit on a box. ⟩ ✓

18.6 Review

Terms and notation introduced in this chapter:

Number theory	*Divisors of 0*	$p \mid a$
Remainder	*Fermat's Little Theorem*	$\text{GCD}(a, b)$
Divisor	*Substitution cipher*	$a \equiv b \pmod{d}$
GCD	*Diffie-Hellman scheme*	$\log_r x$
Euclid's algorithm	*Properties of exponents:*	
Relatively prime numbers	$r^a \cdot r^b = r^{a+b}$	
Diophantine equation	$\left(r^a\right)^b = r^{ab}$	
The fundamental theorem		
of arithmetic	*Discrete logarithm*	
Corollary	*The RSA algorithm*	
Congruence modulo d		

Appendix A. Getting Started with Python

This appendix is online

`www.skylit.com/python/GettingStarted.pdf`

Appendix B. Selected Built-In, Math, and Random Functions

See `https://docs.python.org/3/library/` for complete docs and examples.

`help(obj)`	displays help for a function or module
`input(s)`	displays `s` as a prompt, then reads a line of text, typed in by the user, and returns it as a string
`print(s[, ...])`	prints `s` and subsequent arguments

❖ ❖ ❖

`abs(x)`	returns the absolute value of `x`
`max(a, b)`	returns the larger of the values of `a, b`
`min(a, b)`	returns the smaller of the values of `a, b`
`pow(x, e)`	returns `x**e`
`pow(x, n, p)`	returns `x**n % p`
`divmod(n, d)`	returns `(q, r)` such that $n = qd + r, 0 \leq r < d$
`round(x)`	rounds `x` to an integer
`round(x, d)`	rounds `x` to a floating point number with `d` digits after the decimal point

❖ ❖ ❖

`len(s)`	returns the length of a string, list, or tuple
`sum(lst)`	returns the sum of the numbers from a list or tuple
`max(lst)`	returns the largest element of a list or tuple
`min(lst)`	returns the smallest element of a list or tuple
`range(n)`	generates `0, ..., n-1`, as in: `    for i in range(n):`
`range(m, n)`	generates `m, ..., n-1`
`range(m, n, step)`	generates `m, m + step, m + 2*step, ...`
`zip(s1, s2)`	generates pairs of `s1[i], s2[i]`, as in: `    for i in zip(s1, s2)` or `    list(zip(s1, s2))`

❖ ❖ ❖

Continued ➪

`int(s)`	converts a string or a `float` into an integer
`float(s)`	converts a string or an `int` into a `float`
`str(n)`	converts n into a string
`bin(n)`	returns a string that represents n in binary
`hex(n)`	returns a string that represents n in hex
`oct(n)`	returns a string that represents n in octal

❖   ❖   ❖

`list(s)`	converts a string or tuple into a list
`tuple(s)`	converts a string or list into a tuple
`set(s)`	converts a string or list or tuple into a set
`sorted(s)`	returns a sorted list of elements of s
`reversed(s)`	returns a list of elements of s in reverse order

❖   ❖   ❖

`open(pathname)`	opens a file for reading
`open(pathname, 'w')`	creates a file and opens it for writing

❖   ❖   ❖

`from math import *`

`sqrt(x)`	returns $\sqrt{x}$
`exp(x)`	returns e^x
`log(x), log10(x), log(x, b)`	return $\ln x$, $\log_{10} x$, $\log_b x$, respectively
`sin(x), cos(x), tan(x)`	return the value of the corresponding trig function
`asin(x), acos(x), atan(x)`	return the value of the inverse trig function
`radians(x), degrees(x)`	return x converted to radians, degrees, respectively
`pi`	3.14159...
`e`	2.71828...

❖   ❖   ❖

`from random import *`

`x = random()`	returns a random float $0 \leq x < 1$
`r = randint(a, b)`	returns a random integer $a \leq r \leq b$
`choice(s)`	returns a random element of s
`shuffle(lst)`	shuffles `lst` randomly in place

Appendix C. String Operations and Methods

See https://docs.python.org/3/library/stdtypes.html#text-sequence-type-str for complete docs and examples.

Contents category

The following methods return `True` if all the characters in s belong to the corresponding category; otherwise, they return `False`:	
`s.isalpha()`	Each character in s is a letter (a...z, A...Z)
`s.isdigit()`	Each character in s is a digit (0...9)
`s.isalnum()`	Each character in s is either a letter or a digit
`s.isupper()`	Each letter in s is uppercase
`s.islower()`	Each letter in s is lowercase
`s.isspace()`	Each character in s is "white space" (space, newline, tab, etc., as defined in `string.whitespace`)

Examples:

```
>>> 'ab7'.isalpha()
False
>>> 'ab7'.isdigit()
False
>>> 'a7B'.isalnum()
True
>>> 'AB7'.isupper()
True
```

```
>>> 'a7b'.islower()
True
>>> ' * '.isspace()
False
>>> '  \n\t'.isspace()
True
```

Length and substrings

`len(s)`	Returns the number of characters in s
`ch = s[i]`	Sets ch to the character at index *i* in s
`s2 = s[i:j]`	Sets s2 to the substring from s[i] to s[j-1]

Examples:

```
>>> len('abcd')
4
>>> 'abcd'[1]
'b'
```

```
>>> 'abcd'[1:3]
'bc'
>>> 'abcd'[:3]
'abc'
```

Search

The following methods return an `int`:	
`s.find(sub)`	Returns the index of the first occurrence of sub in s; if sub is not found, returns -1
`s.rfind(sub)`	Returns the index of the last occurrence of sub in s
`s.count(sub)`	Returns the number of times sub occurs in s

`s.find(sub, start, end)`
`s.rfind(sub, start, end)`
`s.count(sub, start, end)`
	The versions with optional arguments start, end look for sub only within the substring of s between start and end-1

Examples:

```
>>> 'never'.find('e')
1
>>> 'never'.find('x')
-1
>>> 'never'.rfind('e')
3
>>> 'never'.count('e')
2
```

```
>>> 'never'.find('ver')
2
>>> 'never'.find('e',2,4)
3
>>> 'never'.rfind('e',1,3)
1
>>> 'never'.find('ver',2,4)
-1
```

Case conversions

The following methods return a new string:	
`s.upper()`	All the letters are converted to uppercase
`s.lower()`	All the letters are converted to lowercase
`s.capitalize()`	If the first character is a letter, it is converted to uppercase

Examples:

```
>>> 'ab7'.upper()
'AB7'
>>> 'Ab7'.lower()
'ab7'
```

```
>>> 'ab7'.capitalize()
'Ab7'
>>> '7ab'.capitalize()
'7ab'
```

Editing and parsing

The following methods return a new string:	
`s.replace(old, new)`	Replaces every occurrence of `old` in `s` with `new`
`s.strip()`	Removes white space at the beginning and at the end of `s`
`s.split(delim)`	Returns a list of substrings separated by occurrences of `delim` in `s`
`s.splitlines()`	Returns a list of lines in `s` — the same as `s.split('\n')`.

Examples:

```
>>> '1^2^3'.replace('^','--')
'--2--3'
>>> ' ab   \n'.strip()
'ab'
```

```
>>> '1, 2, 3'.split(', ')
['1', '2', '3']
```

Formatting

The following methods return a new string:	
`s.format(value,...)`	Formats `value` (or several values) according to the `format` fields in `s`
`s.ljust(w, fill)`	Left-justifies `s` within a string of length `w` and pads it on the right with the `fill` character (`fill` is optional: if not given, `ljust`, `rjust`, and `center` use the space character by default)
`s.rjust(w, fill)`	Right-justifies `s` and pads it on the left with `fill`
`s.center(w, fill)`	Positions `s` in the middle of a string of length `w` and pads it on both sides with `fill`
`s.zfill(w)`	Right-justifies the string and pads it with 0s on the left — the same as `s.rjust(w, '0')`

Examples:

```
>>> '{0:>4s}{1:7.2f}'.format('$',2.5)
'   $   2.50'
>>> '{0:3d} *{1:4.1f}'.format(2, 3)
'  2 * 3.0'
>>> 'ab'.ljust(6, '*')
'ab****'
```

```
>>> 'ab'.rjust(6)
'    ab'
>>> 'ab'.center(6)
'  ab  '
>>> '12'.zfill(4)
'0012'
```

Appendix D. List, Set, and Dictionary Operations and Methods

Lists

Method/Operation	Description
`len(lst)`	Returns the number of elements in `lst`
`x = lst[i]`	Sets `x` to the `i`-th element of `lst`
`lst[i] = x`	Sets the `i`-th element of `lst` to `x`
`del lst[i]`	Deletes the `i`-th element and decrements the indices of the subsequent elements by one
`del lst[i:j]`	Deletes the elements from `i` to `j-1` and adjusts the indices of the subsequent elements
`lst2 = lst[i:j]`	Creates a copy of the specified slice from `lst` and assigns it to `lst2`
`lst2 = lst[:]`	Creates a copy of `lst` and assigns it to `lst2`
`lst.insert(i, x)`	Inserts `x` at index `i`, shifting the subsequent elements to the right by 1
`lst.append(x)`	Appends `x` at the end of `lst`
`lst.pop(i)`	Returns the `i`-th element and removes it from `lst`
`lst.pop()`	Returns the last element and removes it from `lst`
`lst.remove(x)`	Removes the first occurrence of `x` from `lst`; raises an exception if none found
`lst.index(x)`	Returns the index of the first occurrence of `x` in `lst`; raises an exception if none found
`lst.count(x)`	Returns the number of times `x` occurs in `lst`
`lst.reverse()`	Reverses the order of elements in `lst`; returns `None`
`lst.sort()`	Arranges the elements of `lst` in ascending order; returns `None`

Examples:

```
>>> lst = ['A', 'C']
>>> lst
['A', 'C']
>>> lst.insert(1, 'B')
>>> lst
['A', 'B', 'C']
>>> lst.append('A')
>>> lst
['A', 'B', 'C', 'A']
>>> lst.insert(2, 'A')
>>> lst
['A', 'B', 'A', 'C', 'A']
>>> lst.count('A')
3
>>> del lst[2]
>>> lst
['A', 'B', 'C', 'A']
>>> lst.reverse()
>>> lst
['A', 'C', 'B', 'A']
```

```
>>> lst.index('A')
0
>>> lst.remove('A')
>>> lst
['C', 'B', 'A']
>>> lst.sort()
>>> lst
['A', 'B', 'C']
>>> lst.pop(1)
'B'
>>> lst
['A', 'C']
>>> lst.pop()
'C'
>>> lst
['A']
```

See https://docs.python.org/3/library/stdtypes.html#sequence-types-list-tuple-range
for complete docs and examples.

Sets

Method/Operation	Description
`len(s)`	Returns the number of elements in `s`
`s.copy()`	Returns a copy of `s`
`s.add(x)`	Adds `x` to `s`
`s.remove(x)`	Removes `x` from `s`; raises an exception if `x` is not in `s`
`s.discard(x)`	Removes `x` from `s`; has no effect if `x` is not in `s`
`s.pop()`	Removes and returns an arbitrary element from `s`
`s1.issubset(s2)`	Returns `True` if `s1` is a subset of `s2`
`s.update(s2)`	Adds all the elements from a list, tuple, or set `s2` to `s`

Examples:

```
>>> s = {'A', 'C'}
>>> s
{'A', 'C'}
>>> s.add('B')
>>> s
{'A', 'C', 'B'}
>>> s.remove('A')
>>> s
{'C', 'B'}
>>> s.discard('X')
>>> s
{'C', 'B'}
>>> s.pop()
'C'
>>> s
{'B'}
```

```
>>> s2 = set('ABCD')
>>> s2
{'A', 'C', 'B', 'D'}
>>> s.issubset(s2)
True
>>> s.add('X')
>>> s
{'X', 'B'}
>>> s.issubset(s2)
False
>>> s.update(s2)
>>> s
{'A', 'C', 'B', 'D', 'X'}
```

See https://docs.python.org/3/library/stdtypes.html#set for complete docs and examples.

Dictionaries

Method/Operation	Description
`len(d)`	Returns the number of key-value pairs in d
`x = d[k]`	Sets x to the value associated with the key k in d; if k is not in d, raises a `KeyError` exception
`d[k] = x`	If the key k is in d, changes the value associated with k to x; if k is not in d, adds the `k:x` pair to d
`d.get(k)` `d.get(k, dflt)`	The same as `d[k]`, but returns `None` (or the given default value) when k is not in d
`del d[k]`	Deletes the key k and the associated value from d
`k in d`	Returns `True` if k is in d; otherwise, returns `False`
`set(d)` `set(d.keys())`	Returns a set of all the keys in d
`list(d.values())`	Returns a list of all the values in d
`list(d.items())` `set(d.items())`	Return a list or a set of all (key, value) pairs in d
`d.copy()`	Returns a copy of d
`d.update(d2)`	Adds all the key-value pairs from d2 to d

Examples:

```
>>> months = {'Jan': 31, 'Feb': 28}       >>> del months['Mar']
>>> months                                >>> months
{'Jan': 31, 'Feb': 28}                    {'Jan': 31, 'Feb': 28}
>>> months['Feb']                         >>> months[2020] = 12
28                                        >>> months
>>> months['Mar']                         {'Jan': 31, 'Feb': 28, 2020: 12}
KeyError: 'Mar'                           >>> set(months)
>>> months.get('Mar', 30)                 {2020, 'Jan', 'Feb'}
30                                        >>> list(months.values())
>>> months['Mar'] = 31                    [31, 28, 12]
>>> months                                >>> set(months.items())
{'Jan': 31, 'Feb': 28, 'Mar': 31}         {('Feb', 28), (2020, 12), ('Jan', 31)}
```

See https://docs.python.org/3/library/stdtypes.html#mapping-types-dict for complete docs and examples.

Index